W. Leistenschneider · R. Nagel

Atlas of Prostatic Cytology

Techniques and Diagnosis

Foreword by G. Dhom

With 325 Color and Black and White Illustrations
and 27 Tables

Springer-Verlag
Berlin Heidelberg NewYork Tokyo

Prof. Dr. Wolfgang Leistenschneider

Prof. Dr. Reinhard Nagel

Urologische Klinik und Poliklinik
der Freien Universität Berlin
Klinikum Charlottenburg
Spandauer Damm 130
D-1000 Berlin 19

Translated by

Rick Mills and Deirdre Winter

German edition
Praxis der Prostatazytologie

Library of Congress Cataloging in Publication Data. Leistenschneider, W. (Wolfgang), 1943 – Atlas of prostatic cytology. Translation of: Praxis der Prostatazytologie. Bibliography: p. Includes index. 1. Prostate gland – Diseases – Diagnosis – Atlases. 2. Prostate gland – Biopsy, Needle – Atlases. 3. Diagnosis, Cytologic – Atlases. I. Nagel, Reinhard. II. Title. [DNLM: 1. Cytological Technics – atlases. 2. Prostatic Neoplasms – diagnosis – atlases. 3. Prostatitis – diagnosis – atlases. WJ 17 L532p] RC899.L4513 1985 616.6′5 84-26844

ISBN-13: 978-3-642-70112-2 e-ISBN-13: 978-3-642-70110-8

DOI: 10.1007/978-3-642-70110-8

Softcover reprint of the hardcover 1st edition 1985

Reproduction of the figures: Gustav Dreher GmbH, Stuttgart

2122/3130-543210

Foreword

Diagnostic and exfoliative cytology has today achieved a status that few could have envisaged 20 years ago. While exfoliative cytology has long been employed in gynecological diagnosis, new and rewarding spheres have now developed in which cytological diagnosis plays an important role. Exfoliative cytology, for example, is widely employed as an aid in the continuous assessment of urinary tract tumors, and aspiration cytology in the diagnosis of thyroid diseases. This has given rise to a growing demand for pathologists experienced in cytological diagnosis.

However, clinicians with an interest in morphology were often the first to adopt these simple and safe methods, being naturally attracted by the chance to avoid conducting biopsy, which, in the last analysis, is nothing less than a surgical operation.

In this *Atlas of Prostatic Cytology* Leistenschneider and Nagel expertly demonstrate what can be achieved when clinicians skilled in morphology take an interest in cytological methods. Not only do they have direct contact with the patient, but they also profit from immediately being in a position to assess the results of their diagnostic procedures by examining the specimen obtained.

From the technical point of view aspiration biopsy of the prostate is by no means a simple procedure and the difficulties involved would seem to have been underestimated in the initial phase of enthusiastic acclaim. Consequently it has been not unusual for clinicians and pathologists to be disappointed by the high rate of unsatisfactory samples obtained by this method.

It is therefore to be welcomed that the authors have devoted so much space to the technical details in the opening chapter and have specified minimum requirements for specimens to be used in diagnosis. They have also included numerous secondary findings and artefacts. The main objective of aspiration cytology is to confirm prostatic carcinoma when palpation reveals suspicious symptoms. Today histologists and cytologists are in agreement that a carcinoma can be diagnosed with equal reliability on the basis of both a good quality cytological specimen and material obtained by accurate punch biopsy of the tumor focus.

No book on the techniques of prostate cytology is complete without a detailed set of accompanying illustrations. Readers wishing to acquire the skills necessary for the evaluation of cytological preparations will find an extensive array of superb illustrations of possible findings. Naturally, study of the illustrations does not obviate the need for close, patient scrutiny of one's own preparations; on the contrary, it works as an incentive. Considerable talents in visual perception are also necessary if the sometimes very subtle variations are to be recognized. Despite all attempts to objectify visual impressions with the aid of measurement techniques, qualitative assessment by an experienced microscopist will remain an indispensible part of routine diagnosis work for a while to come. However, experience can only be gained by evaluating a large amount of material. The perceptive reader will soon realize what a wealth of experience has gone into the making of this book. It follows that no one who has not had the opportunity to be regularly confronted with a very wide variety of specimens should attempt such a responsible procedure as the diagnosis of carcinomas, especially that of prostatic carcinoma. One of the best ways to achieve this is without doubt to study specimen series together with a "master of the art," a method which should be employed far more frequently.

How much an experienced specialist can learn from the analysis of cytological specimens is shown in the chapters on the grading of prostatic carcinoma and the evaluation of regressive cellular alterations in conservatively managed prostatic carcinoma.

This *Atlas of Prostatic Cytology* will without doubt win over new supporters for this technique. It is to be hoped that it will also contribute toward improving the technical standards of the method.

G. Dhom
Professor and Chairman
Department of Pathology
Universität des Saarlandes
Homburg/Saar

Contents

Introduction

In 1930, FERGUSON became the first to report the possibility of diagnosing carcinoma of the prostate *cytologically* by means of *transperineal aspiration biopsy*. Seven years later he was able to confirm these first results, even in cases in which the clinical findings were negative or inconclusive. Thereafter, however, this ingenious method for providing morphological confirmation of clinically suspected carcinoma lapsed into obscurity for more than 20 years. This is all the more remarkable in consideration of the fact that Ferguson was able to aspirate diagnosable material in more than 80% of the needle punctures, that the diagnosis was true-positive in more than 70% of cases, and that in *none* of 23 cases did his cytologically experienced pathologist derive a false-negative diagnosis from Ferguson's preparations, as the further clinical course clearly proved.

In the light of the present-day accuracy of cytological analysis of prostatic aspirates, the results of which are comparable to those achieved by histological analysis, it is incomprehensible why this pioneering work was ignored for so long – all the more so in view of the fact that FERGUSON saw the significance of aspiration biopsy as residing exclusively in its capacity to establish the diagnosis of carcinoma and expressly denied that it could be employed to classify or grade a tumor, for which purpose it is now most successfully employed.

With the increasing use of *exfoliative cytodiagnosis* of bladder carcinoma based on the work of PAPANICOLAOU (1954) and others, the possibility of also being able to diagnose carcinoma of the prostate by means of *exfoliative cytology* subsequent to prostatic massage appeared extremely attractive, and for a time, indeed, this technique was frequently used. Soon, however, it was recognized that only in 40–50% of cases could carcinoma of the prostate be established cytologically in prostatic secretion or in urine following prostatic massage (MULHOLLAND 1931; FRANK and SCOTT 1958; GARRETT and JASSIE 1976). Since prostatic massage was also often rejected in the presence of suspect findings or when palpation provided an apparently definite diagnosis of carcinoma, the method ultimately never attained clinical significance.

The *cytological diagnosis of inflammatory diseases* of the prostate by analysis of secretion obtained from middle-aged patients following prostatic massage also proved unreliable, since not even in 50% of cases could the clinical diagnosis of chronic prostatitis be confirmed cytologically (O'SHAUGHNESSY et al. 1956; BOURNE and FRISHETTE 1967; MEARES and STAMEY 1972).

After exfoliative cytology had proven ineffective for the morphological confirmation of the clinical diagnosis for both malignant and inflammatory diseases, FRANZÉN et al. returned to the method of transperineal aspiration biopsy described by FERGUSON 30 years before and reinvestigated its usefulness in the *cytodiagnosis of carcinoma.*

The first attempts to establish so-called early carcinoma by transperineal biopsy met with little success, so FRANZÉN et al. chose the *transrectal route* and, within just a few years, managed to develop the aspiration technique to such perfection that in 1960 they were already able to report about 100 aspiration biopsies without complications.

Aspiration biopsy, which has few complications, causes the patient hardly any discomfort, and is just as efficient as the histological assessment of prostatic diseases, has until today – apart from recent initial trials in the United States (KAUFMAN et al. 1982; MELOGRANA et al. 1982) – found widespread use almost solely in Europe. Nevertheless, the validity of aspiration biopsy in both the primary diagnosis of prostatic carcinoma and secondary prostatic tumors and the diagnosis and classification of inflammatory diseases of the prostate is as undisputed as its reliability in the evaluation of the effect of therapy in the locally advanced, conservatively treated carcinoma of the prostate, thus fulfilling expectations which 50 years ago Ferguson emphatically rejected as being beyond its means.

During the early days of cytodiagnosis of prostatic carcinoma, doubts were repeatedly cast on the technique's reliability and on the possibility of deriving sound therapeutic decisions from analysis of the aspirate. These doubts have proven groundless, as long as technically perfect aspirate is obtained, it is properly prepared, and an experienced cytologist performs the evaluation. Under these circumstances a definite cytological diagnosis of prostatic carcinoma no longer requires 'confirmation' through histological analysis of a specimen collected by *punch biopsy*, since the rate of false-positives in the primary diagnosis of prostatic carcinoma is *not* higher in cytological than in histological investigations. However, despite its low complication rate, aspiration biopsy is *not* suited for use as a method of screening for certain gynecological tumors for example.

But despite the substantial advantages it has over punch biopsy, aspiration biopsy has not yet found equally wide application. This is explained by the fact that it is far more difficult to learn the aspiration technique than the technique of punch biopsy. Not only does it require of the urologist a mastery of the aspiration technique itself but it also necessitates correct technical processing – smearing and fixation – of the cellular material in order that a qualitatively perfect preparation be obtained which permits a reliable cytological evaluation.

These difficulties, which can easily be overcome by practice, are far outweighed by the many advantages of the method for the patient and the urologist. There therefore seems to be sufficient justification for a detailed account of the possibilities of cytology for the diagnosis, treatment, and treatment control of *all* prostatic diseases, together with a description of the technical background of aspiration, fixation, and staining procedures.

Methodologically, this book is based on the evaluation of preparations which were obtained by transrectal aspiration biopsy, fixed wet, and stained according to Papanicolaou; particularly characteristic findings were additionally stained with May-Grünwald-Giemsa solution.

1 The Technical Bases of Aspiration Biopsy

The quality of a cytological preparation, and thus the possibility of evaluating it, depends almost solely on:

- mastery of the puncture technique, and
- meticulous processing (smearing, fixation, staining) of the aspirated cellular material.

Given these preconditions, reliable and reproducible assessment of the aspirate by the cytologist is possible in almost all cases.

In the literature, the proportion of specimens that *cannot* be evaluated as a result of either quantitative or qualitative deficiencies in cellular material is cited as 0.7–60% **(Table 1).** The figure is lowest in centers or practices that have the necessary extensive experience in the aspiration and processing of the biopsy material (ESPOSTI 1966, 1975; FAUL 1975; ACKERMANN and MÜLLER 1977; LEISTENSCHNEIDER and NAGEL 1980), and generally increases among specimens sent to a cytodiagnostic institute in those cases in which the aspiration biopsy is seldom carried out by the urologist submitting the specimens (HOHBACH and DHOM 1978; FAUL 1980).

Table 1. Aspiration biopsy according to FRANZÉN: frequency of inadequate cellular material

Author(s)	Year	Biopsies	
		n	%
ESPOSTI	1966	1430	2.0
ESPOSTI	1975	345	1.2
FAUL	1975	1382	3.0
DROESE et al.	1976	288	6.4
BISHOP and OLIVER	1977	182	16.4
ACKERMANN and MÜLLER	1977	645	6.4
HOHBACH and DHOM	1978	100	60.0
FAUL	1980	415	53.0
ESPOSTI	1982	4630	0.7

1.1 Preparation and Positioning of the Patient

The principle of transrectal aspiration biopsy is depicted in **Fig. 1.**

Special preparation of the patient is unnecessary; nevertheless, evacuation should have taken place prior to the biopsy in order that the ampulla recti be free of stool.

By contrast, the *positioning of the patient* is of considerable importance since it can exert a vital influence upon the puncture.

In the *lithotomy position,* the nates should project almost a hand's width beyond the table edge and the examination table should be raised to its highest possible position, as only then can the needle be accurately inserted into the suspect area of the prostate **(Fig. 2).** If the table is too low, the direction of puncture will almost inevitably be tangential; as a result the prostate will merely be grazed by the needle, with the consequence that the suspect area is easily missed or an insufficient amount of cellular material is aspirated.

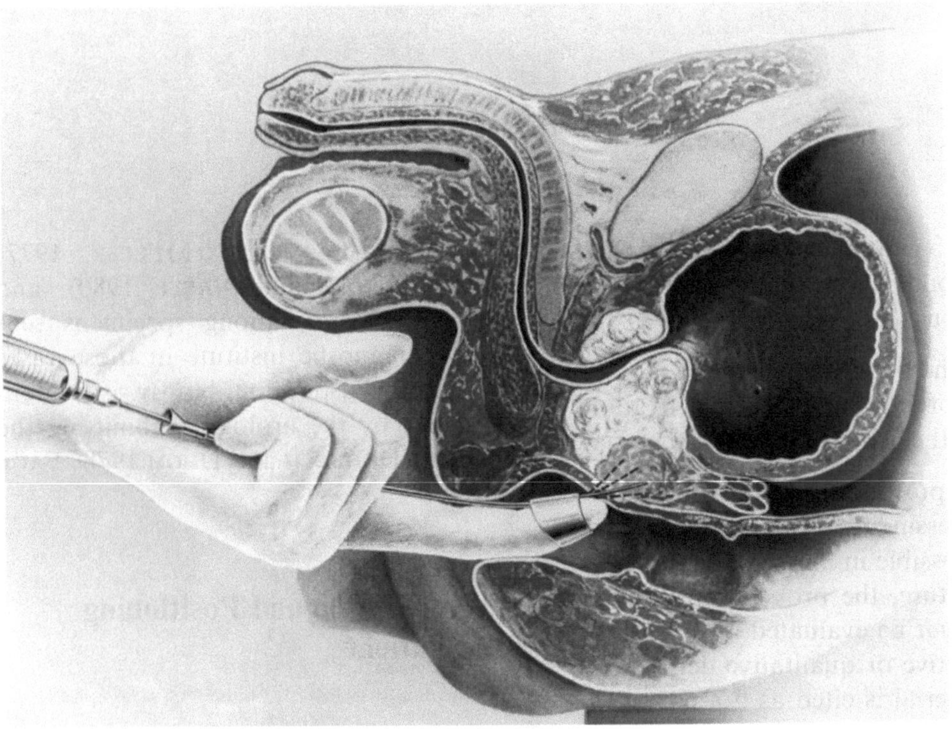

Fig. 1. Principle of aspiration biopsy of the prostate according to FRANZÉN

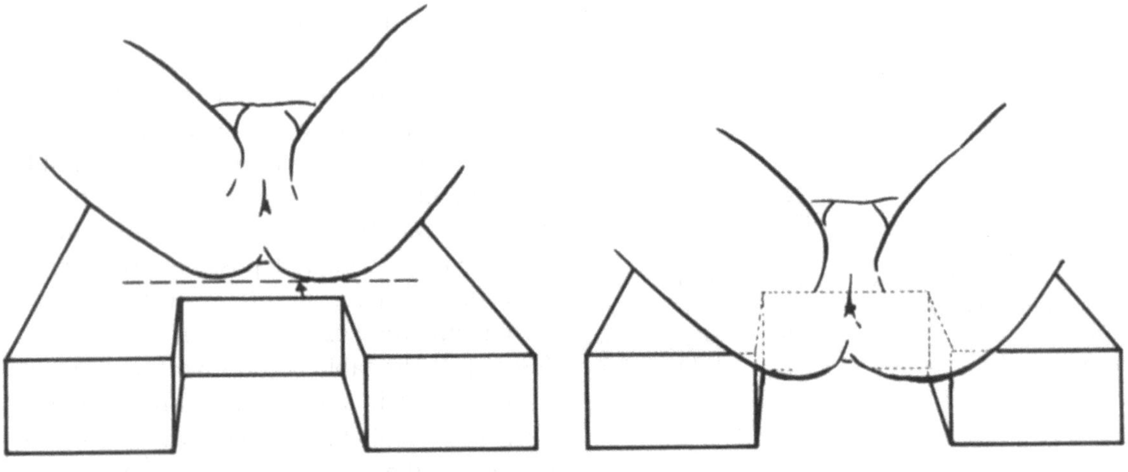

Fig. 2. Positioning of the patient for aspiration biopsy. *Left:* wrong! *Right:* correct! The patient's nates extend approximately a hand's width beyond the edge of the examination table

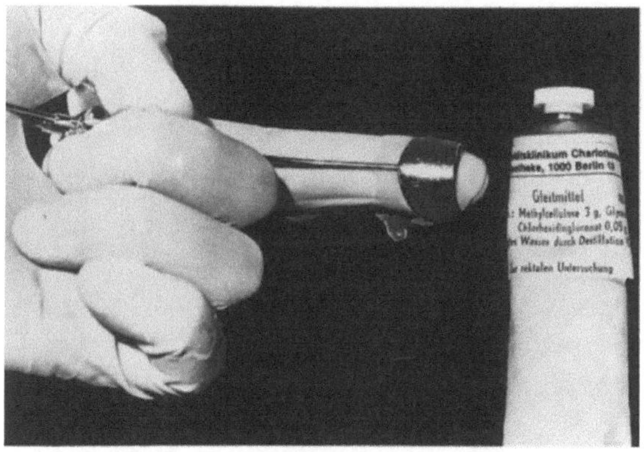

Fig. 3. Prior to introduction of the needle guide, ample lubricant is applied to the tip of the index finger

1.2 Lubricants and Anesthesia

A great advantage of aspiration biopsy is the exceedingly low level of pain which it causes in comparison with other biopsy techniques, and local anesthesia is therefore very rarely necessary. In combination with the very low complication rate (see p. 14) this advantage makes it possible, even in a urological practice, not only to perform repeat aspirations when palpation continues to evoke suspicion despite negative cytological findings, but also to carry out short-term control biopsies for objective assessment of the therapeutic effects on the primary tumor itself after certain periods of treatment (see p. 107).

1.2.1 Lubricants

As with every rectal examination, prior to aspiration biopsy so much lubricant is applied to the index finger that it begins to drip **(Fig. 3).** This permits easy introduction of the finger together with the metal steering ring at the end of the needle guide and the needle guide itself.

The lubricant should be *watery,* since ointments (Vaseline, boric acid ointment etc.) cause impurities (artefacts) in stained specimens which can severely impair diagnosis.

In our experience a lubricant composed of the following constituents has proven satisfactory:

chlorhexidine gluconate	0.05
methylcellulose	3.0
glycerol	5.0
distilled water	92.0

1.2.2 Local Anesthesia

In those patients who have already found a normal rectal examination uncomfortable, instillation of lubricant containing a topical anesthetic (e.g. 11 ml Instillagel) into the ampulla recti is to be recommended. After some 5–10 minutes the mucous membrane will be adequately anesthetized, and the aspiration biopsy can be carried out without causing the patient pain.

If, despite this precaution, the patient reports pain when the finger and needle guide are inserted or reacts with involuntary con-

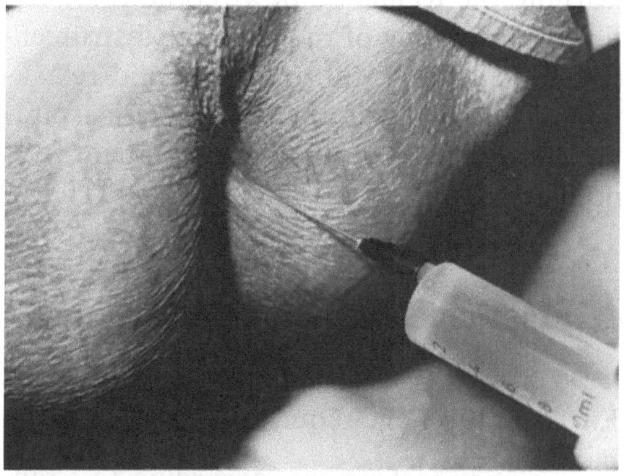

Fig. 4. Local anesthesia in the area of the m. sphincter externus: the local anesthetic is injected at 3 o'clock and from that point in a semicircle towards 12 and 6 o'clock. The same procedure is performed on the opposite side

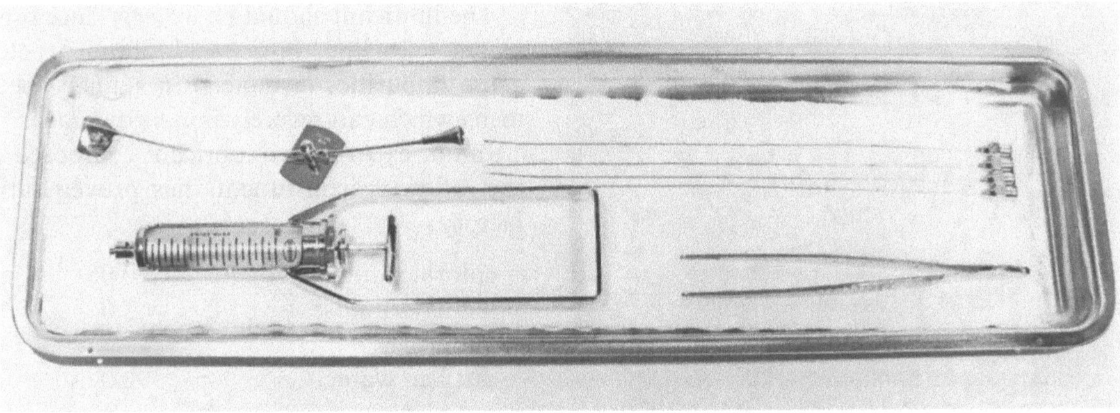

Fig. 5. Complete instrumentarium for aspiration biopsy, with four needles and forceps alongside the basic equipment

traction of the sphincter ani, infiltration of the m. sphincter ani externus with 10 ml of a 1% local anesthetic is indicated. This has previously proven its efficacy during punch biopsy (LEISTENSCHNEIDER and NAGEL 1978).

Using a thin needle, the m. sphincter ani is infiltrated circularly with 5 ml of the anesthetic at each of two puncture sites (3 o'clock and 9 o'clock) **(Fig. 4).**

The use of local anesthesia is especially recommended during the initial phases of learning the technique of aspiration. At this time the handling of the puncture instrumentarium is usually still uncertain, and relaxation of the sphincter ani will facilitate aspiration as well as ensuring that the patient feels no pain.

1.3 Instrumentarium for Aspiration

For aspiration one requires:

- aspiration instruments according to Franzén (syringe, needles, needle guide)
- small anatomical forceps
- sterile gloves
- microscopic slides
- diamond marking pencil for inscribing the slides

The *sterile puncture instrumentarium* for one patient should contain four needles **(Fig. 5),** since in most cases up to four aspirates – two per lobe – are necessary to procure sufficient cellular material.

After each individual aspiration a new needle must be used, since expression of the aspirate onto the slide renders the needle unsterile.

Aspiration should not be performed with even *slightly bent needles* since the removal of diagnosable material is no longer assured.

Since aspiration of cells into the needle requires a high *vacuum* in the syringe, the syringe should be checked for tightness prior to aspiration. This is done, after donning the surgical gloves, by withdrawing the plunger as far as possible while simultaneously covering the nozzle of the syringe with the finger. Inability to create a vacuum either *before* or *during* aspiration may be due to the following factors:

- syringe not tightly screwed together
- wear upon the glass cylinder of the syringe
- a gap between the needle and the syringe

1.4 Technique of Aspiration

The needle guide is threaded over the left index finger (or over the right index finger if one is left handed) in such a way that the adjustable metal plate can be fixed on the thenar using the middle or ring finger (guitarist's grip), with the result that the steering ring at the end of the guide is secured around the tip of the index finger and cannot slip **(Fig. 6).**

The index finger, moistened with ample lubricant, and the needle guide are now introduced into the rectum. Simultaneously the patient is asked to 'push', as if attempting to defecate.

The tip of the finger, together with the steering ring, is then advanced to the suspect area of the prostate. It is of decisive importance that the ring does not ride up the finger **(Fig. 7),** since direct contact with the suspect area will no longer be possible.

After withdrawing the plunger as far as level 1 (=1 ml air), the thin aspiration needle, which has previously been connected to the syringe, is introduced cautiously into the guide in a straight line so that it does not bend **(Fig. 8).**

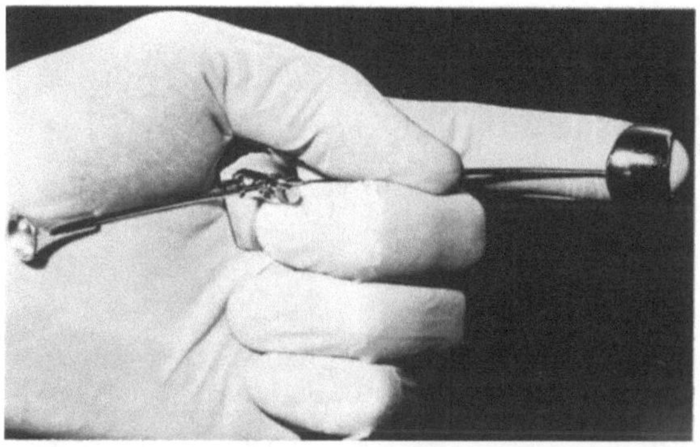

Fig. 6. Correct position of the steering ring at the end of the needle guide; during both the introduction and the biopsy procedure itself the ring must always be placed at the tip of the index finger

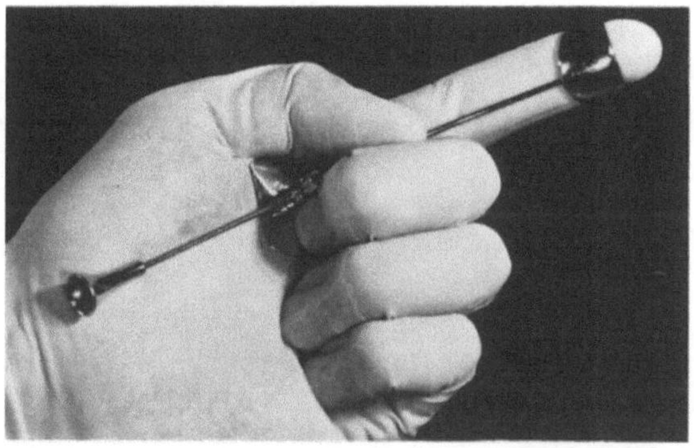

Fig. 7. Incorrect positioning of the steering ring, so that the tip of the index finger projects too far

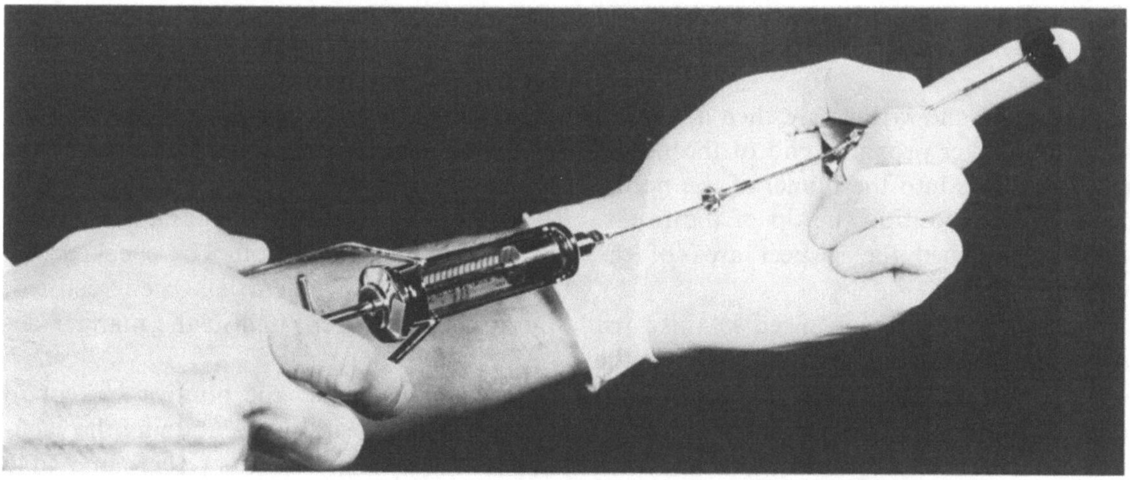

Fig. 8. Correct position during introduction of the aspiration needle into the needle guide

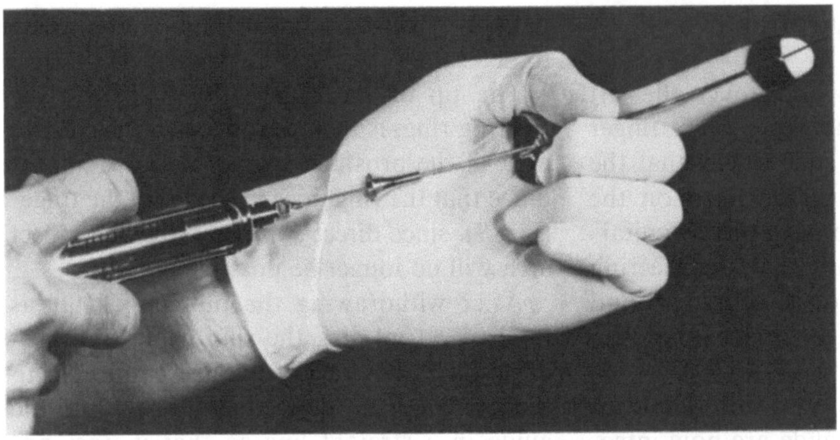

Fig. 9. The aspiration needle is advanced, and in this position penetrates the suspect area of the prostate. The right hand is then used to inject 1 ml air forcibly into the prostate

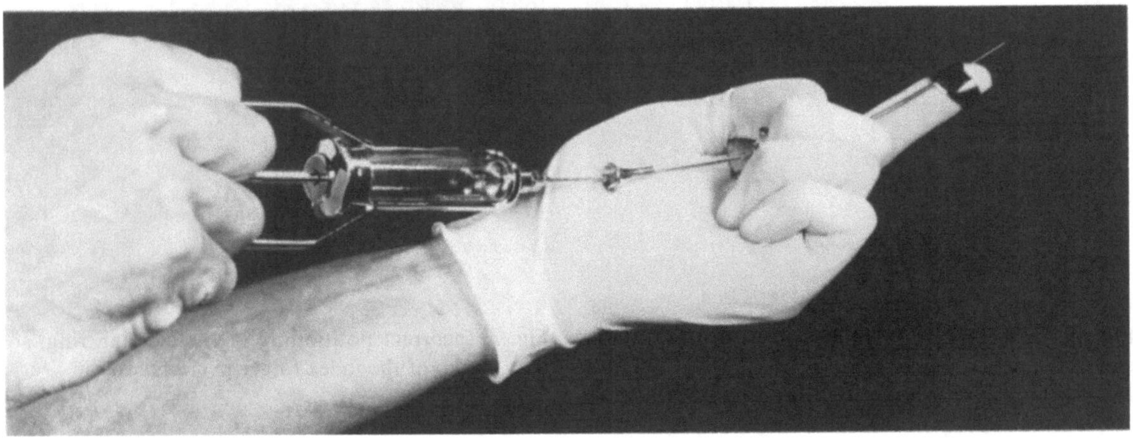

Fig. 10. Final position of the instruments for aspiration: the plunger is fully withdrawn, thus creating the suction necessary for the biopsy

The needle and syringe are then advanced until the thicker proximal end of the needle just disappears into the funnel of the needle guide. In this position the tip of the needle will have reached the suspect area of the prostate **(Fig. 9).**

The needle is now advanced in this area to a depth of about 1 cm, after which the previously drawn-in air is *rapidly expelled* in order to loosen sheets of prostatic epithelial cells from the basement membrane and thus enable sufficient cellular material to be aspirated. The plunger is then pulled back forcibly, thus creating the vacuum necessary for aspiration **(Fig. 10).**

If this *fails* to create a vacuum, the sources of error cited above should be checked (see p. 6). After the problem has been rectified, the procedure is repeated in the manner described.

During the biopsy the plunger is kept in its withdrawn position, maintaining the vacuum in the syringe, while at the same time the needle is moved back and forth *rapidly*

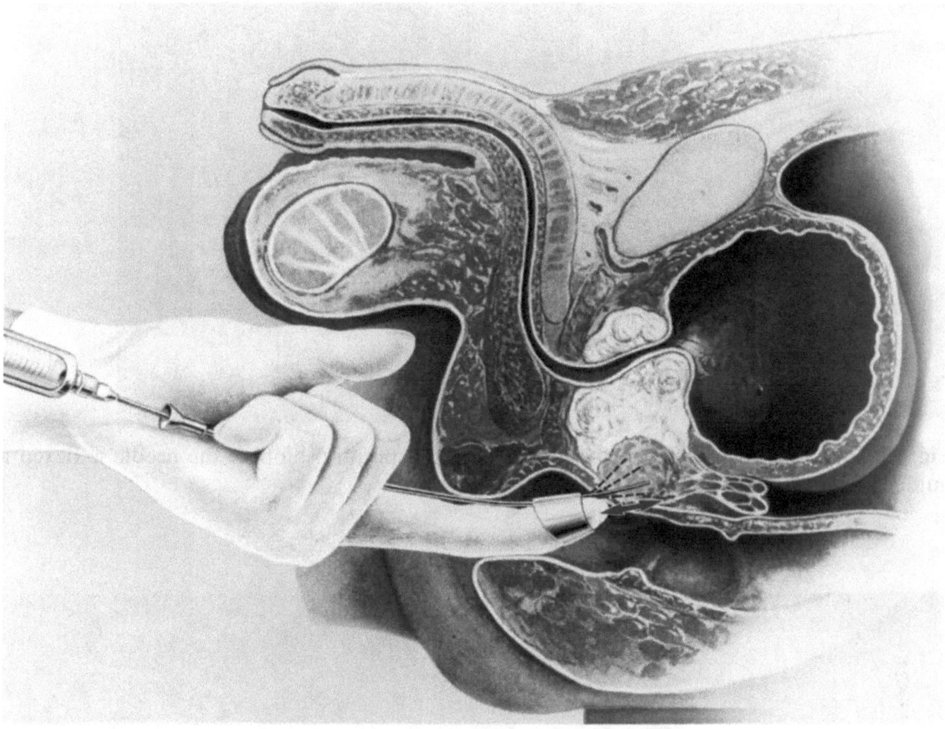

Fig. 11. In this position the tip of the needle is moved rapidly back and forth (10–15 × !) within the suspect area in order to procure cellular material

10–15 times in a *fan-like fashion* at a depth of 1 cm within the suspect area. The quicker the movement of the needle within the prostate, the less painful is the biopsy **(Fig. 11).**

This is technically the most difficult part of the biopsy, since on no account may the needle be allowed to slip out of the prostate during these movements. Should this happen, the vacuum would immediately be lost.

This phase of the procedure is considerably facilitated by resting the elbow of the hand guiding the needle on the hip **(Fig. 12).**

During the rapid movement of the needle care must be taken that it does not under any circumstances flex against the needle guide **(Fig. 13).**

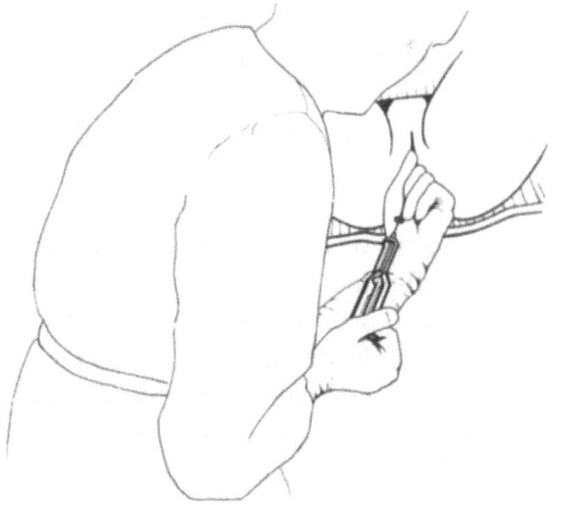

Fig. 12. Optimal stance of the operator, the elbow being supported by the hip during aspiration. In this way the rapid movement of the needle within the suspect area of the prostate is rendered considerably less hazardous

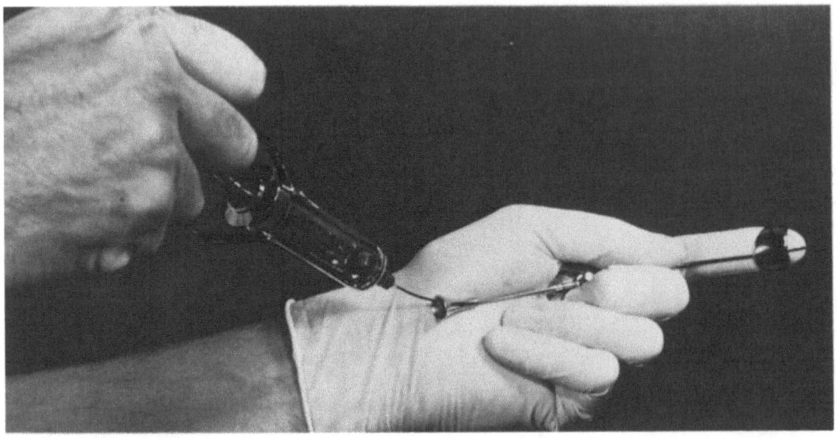

Fig. 13. Incorrect positioning of the needle during aspiration biopsy: the needle is flexed against the needle guide

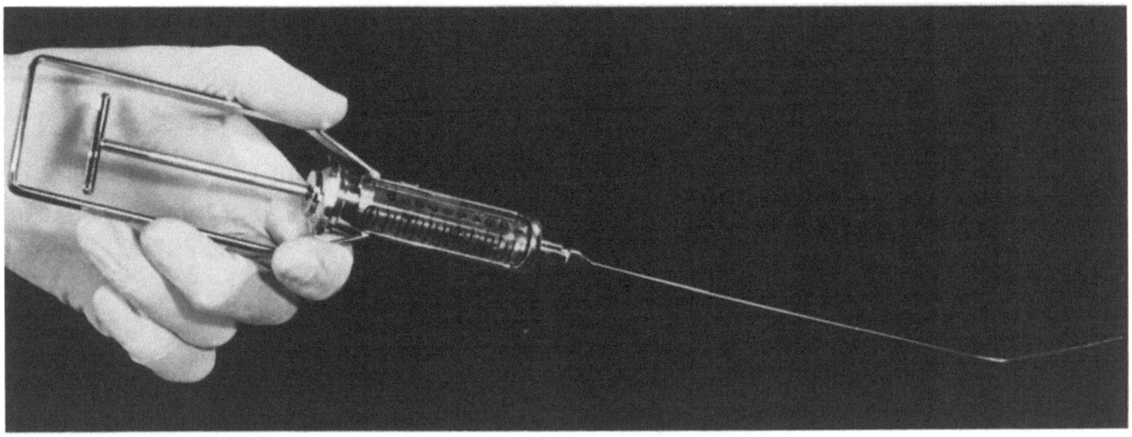

Fig. 14. Bending of the aspiration needle, making it useless for biopsy

When the tip of the needle is found to be badly bent after the aspiration, this is usually a sign that the prostate has been pierced in the direction of the pelvic wall **(Fig. 14).** Consequently, adequate cellular material cannot be expected and the procedure must be repeated with a new needle.

Prior to withdrawing the needle from the prostate it is vital that the plunger be released, so that it returns to its original position as a result of the still existing vacuum.

If one withdraws the needle from the prostate while the plunger is still pulled back, the aspirated cellular material will be *sucked into the syringe* by the previously formed vacuum. In this way the aspirate will be lost to diagnosis, since it can no longer be expressed onto the slide.

If the plunger of the syringe does not return to its original position at the end of the biopsy, although the needle is still in the prostate, a loss of vacuum has occurred during the procedure.

After the plunger has returned to its original position, the entire aspiration equipment is removed from the rectum; during its removal, the patient again 'presses' as if defecating so that the finger and steering ring can be withdrawn without inflicting pain.

After rapidly disconnecting the needle – if necessary with the help of the sterile forceps already laid out for the purpose – the plunger is withdrawn fully once more and the needle again connected tightly onto the syringe. By depressing the plunger as *rapidly and completely* as possible, the aspirate is expressed out of the needle and onto a slide which has previously been numbered and inscribed with the patient's initials. The material should rest approximately 1 cm from the edge of the slide **(Fig. 15)**.

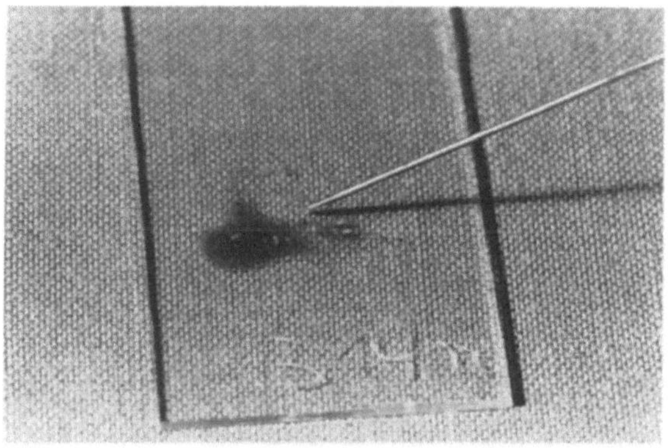

Fig. 15. The aspirate is expressed onto one end of a clean slide, the edge of which has been engraved with the registration number using a diamond marking pencil

1.5 Acquisition of Suitable Cellular Material

An essential precondition for reliable cytological diagnosis is the presence of sufficient sheets of prostatic epithelial cells in the aspirate.

At least 20 medium sized, well preserved sheets are required from each lobe; any less, and reliable diagnosis will not be possible.

While the beginner must always procure three aspirates from each suspect area, using a new needle on every occasion, even the experienced operator must obtain at least two aspirates. By contrast, three aspirates should always be taken from small suspected foci to ensure that nothing is overlooked. It is also advisable to obtain two aspirates from the lobe in which palpation is not suggestive of carcinoma, since multifocal growth is present in approximately 60% of cases (Byar and Mostofi 1972; Kastendieck et al. 1976).

1.6 Macroscopic Evaluation of the Aspirate in Biopsy Smears

1.6.1 Satisfactory Aspirate

The typical prostatic aspirate is grayish white and somewhat glutinous; small amounts of blood in the aspirate are tolerable. A very abundant, often sticky, homogeneous grayish white aspirate is typical for a locally advanced carcinoma.

1.6.2 Unsatisfactory Aspirate

The presence of seminal vesicle secretion or urine in the aspirate may render it quantitatively or qualitatively inadequate, while fine fibres in the smear indicate contamination by epithelial cells from the rectal mucosa.

1.7 Smear Technique

After the aspirate has been expressed onto a slide, smearing is achieved by pressing a second slide, at right angles to the first, down *evenly* upon the aspirate with moderate pressure (beware of breaking the glass) so that the aspirate is spread across the one side of the slide **(Fig. 16).**

The upper slide is now moved evenly towards the narrow edge of the lower slide lying opposite to the aspirate, smearing the aspirated material regularly in a thin layer across the slide **(Fig. 17).**

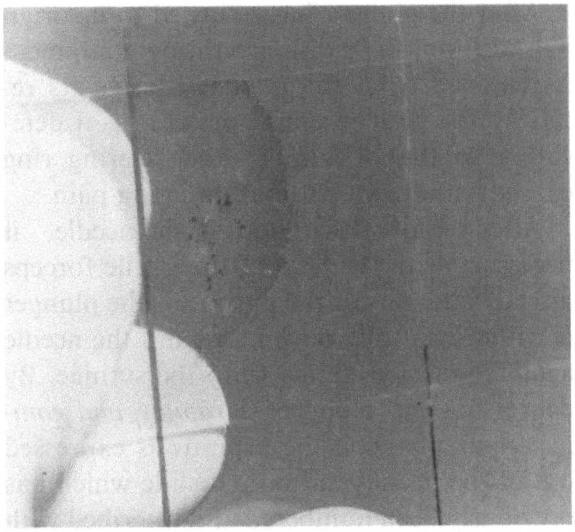

Fig. 16. The aspirate is first evenly distributed by pressing a second slide down flat upon it

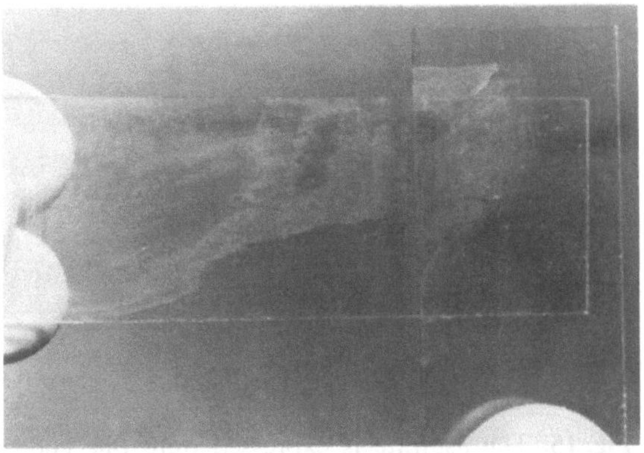

Fig. 17. Under continuous slight pressure, the upper slide is moved to the right, across the lower slide, evenly smearing the aspirate

1.8 Fixation

Three methods of fixation have proven efficient:

- spray
- alcohol-ether solution
- air drying

The method to be employed must be discussed with the cytodiagnostician, since this will depend upon the staining procedure.

1.8.1 Spray

When performed correctly, fixation of the aspirate using a commercially available spray containing polyethylene glycol (Merckofix®) is equally good a method of 'wet fixation' as fixation by means of alcohol-ether solution. Nevertheless, it is more difficult to employ, and optimal results will be achieved only when the following conditions are strictly complied with:

- **Fixation must occur within 5 seconds of smearing,** since – as with every wet fixation technique – even slight drying of the cells through contact with the air will lead to changes that hinder or render impossible their evaluation subsequent to staining.
- The smear must be sprayed for 4–5 seconds from a distance of 20–25 cm. During

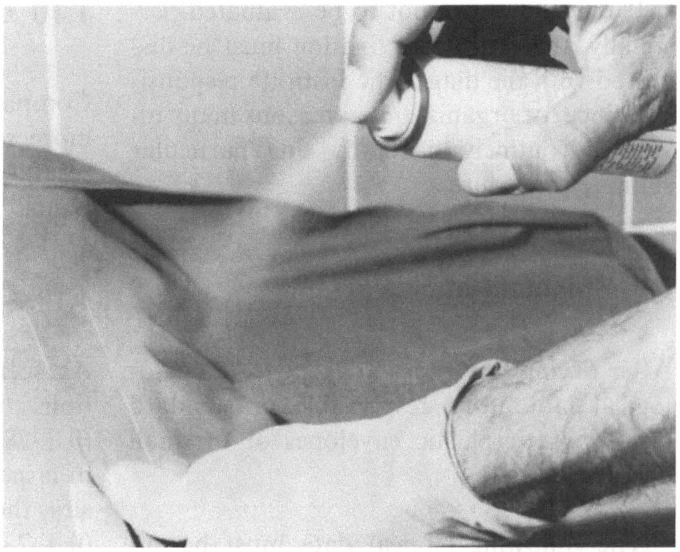

Fig. 18. Within a maximum of 5 s after smearing (!), the fixative is sprayed evenly upon the slide for 4–5 s from a distance of 20–25 cm

this time the spray must be moved quickly to and fro horizontally above the specimen, the spray descending vertically upon the smear **(Fig. 18).**

If the spray is held closer to the preparation than 20 cm, there is a danger that the cellular material will be partially or totally destroyed through freezing; if the spray is held further away, partial or total autolysis may occur.

- The smear must be allowed to dry for at least 15 minutes, after which it can be stained or stored indefinitely.

Provided that these guidelines are strictly adhered to, spray fixation has proven to be the most practicable and reliable method of fixation, especially in the urological practice.

1.8.2 Alcohol-Ether Solution

A mixture of 96% alcohol and ether in equal parts or 99% isopropyl alcohol has proved effective for fixation (Soost 1978). A certain disadvantage, however, is the rapid evaporation of the ether. Containers must bear the inscription "no naked lights".

Because of the problematic nature of these solutions and the difficulties involved in their use, it seems to us that they are less suitable for the urological practice, especially when the smears have to be sent away for analysis.

Again, with this method of fixation it is crucial that the smears be plunged into the solution immediately after they have been prepared.

1.8.3 Air Drying

An alternative to wet fixation is fixation through simple air drying. If this method is used, however, only staining after May-Grünwald-Giemsa or with hematoxylin-eosin is subsequently possible. Fixation of smears by air drying is completed after 60 minutes.

If the smears are not to be evaluated personally, the method of fixation must be discussed with the diagnostic institute responsible, since for organizational reasons many institutes routinely use only one particular staining procedure.

1.9 Shipment of Smears

After drying, the slides are sent to the cytological laboratory responsible in standard cardboard or plastic envelopes or in small wooden cases.

The following clinical data must be enclosed for every patient:

- surname, Christian name, date of birth
- date of aspiration, including the year
- number of aspirates per prostatic lobe
- method of fixation

For primary biopsy:
- questions arising from clinical examination

For therapeutic control:
- form and duration of treatment

1.10 Complications

Complications after aspiration biopsy are far more seldom and above all usually far less serious than after transrectal and perineal punch biopsy.

1.10.1 Complications of Punch Biopsy

As seen in representative series, complications must be expected in about 8–10% (0.3–28%) of punch biopsies **(Table 2),** while in large series of transrectal aspiration biopsies the complication rate averages 1.7% (0.4–2.4%) (p. 16).

The most frequent complications after either punch or aspiration biopsy are:

attacks of fever
bleeding from the puncture channel
bleeding from the rectum (hemorrhoids)
gross hematuria
epididymitis
prostatitis
hemospermia
septicemia

Table 2. Complications of transrectal and transperineal prostatic punch biopsies in representative series

Transrectal prostatic punch biopsies				Transperineal prostatic punch biopsies			
Author(s)	Year	No. of biopsies	Complications (%)	Author(s)	Year	No. of biopsies	Complications (%)
DAVES et al.	1961	175	1.7	PECK	1960	164	2.4
EMMETT et al.	1962	203	~3	KAUFMAN and SCHULTZ	1962	704	2.2
ROTHKOPF	1966	116	4.3	ARDUINO and MURPHY	1963	171	24.8
MAKSIMOVIĆ et al.	1971	148	10.8	MAKSIMOVIĆ et al.	1971	121	8.3
TÜMMERS and WEISSBACH	1975	283	17	HELL et al.	1971	214	2.3
KÖLLERMANN et al.	1975	535	6.7	CHIARI and HARZMANN	1975	123	17.8
CHIARI and HARZMANN	1975	131	28.1	PUIGVERT et al.	1975	1500	0.3
LEISTENSCHNEIDER and NAGEL	1978	977	11.9				

Table 3. Complications in a personal series of 977 transrectal punch biopsies (11.9%) with the TRU-CUT needle

Complication	Mild	Severe
Fever	14 (1.4%)	–
Gross hematuria	29 (2.9%)	9 (0.9%)
Bleeding from the rectum/ puncture channel	35 (3.6%)	17 (1.7%)
Epididymitis	–	12 (1.2%)
Prostatitis	–	1 (0.1%)
	78 (7.9%)	39 (4%)

In general these complications are classified as 'severe' when active treatment (e.g. high doses of antibiotics, surgical hemostasis by ligation) is necessitated or when the complication per se is serious (epididymitis, prostatitis, septicemia).

The frequency of **severe complications** after perineal and transrectal punch biopsy is similar, the reported rates being 1.8–2.6% for *perineal biopsy* (Peck 1960; Kaufman and Schultz 1962; Arduino and Murphy 1963) and 0.9–3.98% for *transrectal biopsy* (Chiari and Harzmann 1974; Köllermann et al. 1975; Leistenschneider and Nagel 1978).

In a *personal series* of 977 transrectal punch biopsies, mild complications occurred in 7.9% of cases and severe complications in 4%, though in no instance were the latter fatal **(Table 3).** In all patients two or three punches were made per biopsy.

To date, no more than five deaths have been published that were regarded as a direct or indirect consequence of the biopsy (Bertelsen 1966; Wendell and Evans 1967; Davison and Malament 1971).

1.10.2 Complications of Aspiration Biopsy

In 1076 aspiration biopsies, none of which was performed simultaneously with punch biopsy, we have observed only 21 complications (1.9%), of which 0.49% were considered severe (epididymitis, colibacillemia).

	$n=21$	%
Fever	12	1.1
Bleeding from the rectum	2	0.19
Hemospermia	2	0.19
Epididymitis	4	0.4
Colibacillemia	1	0.09

Our personal observations show that the incidence of *fever* is almost as high as after transrectal punch biopsy, whereas *epididymitis* is three times less frequent and *rectal bleeding* 28 times less frequent **(Table 3).**

Severe local hemorrhagic complications after aspiration biopsy have not been encountered to date, and in view of the thin needle used are scarcely to be expected.

In other large series too, complications after aspiration biopsy are, at an average of 1.4% (0.4–2.4%), remarkably infrequent in comparison with those after punch biopsy; this is evident from **Table 4,** which contains information on 8190 biopsies.

During the therapeutic control of conservatively treated prostatic carcinoma (six different forms of treatment) we observed complications in only 1.6% of the 498 aspiration biopsies performed *without* simultaneous punch biopsy (out of a total of 600 aspiration biopsies).

Esposti and co-workers, who as early as 1975 reported on more than 14000 aspiration biopsies, observed a total complication rate of less than 1% and only four cases of septicemia. The same authors established that in the presence of chronic prostatitis with *E. coli* infection of the urine, the complication rate is significantly higher (1.5%). And in a small group of 32 patients with chronic polyarthritis, complications arose following as many as 7.1% of the 42 biopsies performed.

Table 4. Complications of transrectal aspiration biopsy of the prostate according to Franzén

Author(s)	Year	No. of biop-sies	Compli-cations
Esposti et al.	1968	3002	12 (0.4%)
Faul	1975	2336	42 (1.8%)
Staehler et al.	1975	1020	24 (2.4%)
Esposti et al.	1975	571	9 (1.6%)
Köllermann et al.	1975	185	4 (2.2%)
Leistenschneider and Nagel	1982	1076	21 (1.9%)

In cases of acute prostatitis, aspiration biopsy is clearly contraindicated, particularly since it is of no relevance for the clinical diagnosis (Esposti et al. 1975; Faul 1975; Leistenschneider and Nagel 1978).

By contrast, and despite the somewhat increased risk of complications, aspiration is indicated in *chronic prostatitis* in the presence of suspect findings at palpation and after careful consideration of differential diagnosis. The reason for this is the known fact that in 10% of these patients a prostatic carcinoma can be detected cytologically. Differentation between granulomatous prostatitis and carcinoma, especially is often impossible clinically, and it can even be difficult morphologically.

1.11 Prophylaxis Against Infection

A uniform approach to the prevention of inflammatory complications of aspiration biopsy is not to be found in the literature. It is, however, certain that antibiotics or chemotherapy *prior* to aspiration serve no purpose.

After the biopsy we submit all patients to a 5-day course of chemotherapy (sulfoxazole in combination with trimethoprim) at standard dosages. However, the patients, who are generally elderly, must continually be reminded to take the tablets in accordance with instructions.

Despite this chemotherapeutic prophylaxis, the infection-related complication rate (1.49%) is not significantly lower than in other large series.

All of our patients who experience an infection-related complication after aspiration biopsy receive a broad-spectrum antibiotic effective against Gram-negative bacteria over 5 days subsequent to repeat biopsy. We have made this routine procedure since in three patients who ran high temperatures after the initial biopsy and in whom a recurrence of high fever after follow-up biopsy was not controlled by repeat administration of the same chemotherapeutic agent, the symptoms subsided following treatment with an appropriate antibiotic.

Although infections are a very rare complication after aspiration biopsy, patients should be instructed to return or go to hospital immediately (even at night!) if they develop fever or attacks of shivering, so that intensive antibiotic treatment can be commenced at once.

1.12 Staining Procedures

Most prostatic aspirates are stained with solutions according to Papanicolaou or May-Grünwald-Giemsa (MGG). In addition, hematoxylin-eosin (H&E) is sometimes used.

While staining according to May-Grünwald-Giemsa can only be performed on *air-dried smears,* Papanicolaou staining is carried out on wet-fixed smears (those fixed with spray or alcohol-ether solution). Hematoxylin-eosin staining can be performed on both wet-fixed and air-dried smears; nevertheless, optimal results are obtained after wet fixation.

1.12.1 Staining According to Papanicolaou

The Papanicolaou method of staining provides the greatest clarity of detail (TAKAHASHI 1981; ESPOSTI 1982). In addition, the reproducibility of the findings is excellent, which is very important both for primary and regression grading, and for the classification of inflammatory changes.

Characteristic for this stain are the pronounced transparency and excellent color differentiation of the cytoplasm that result from the polychromatic quality of the staining solution, with its cationic, anionic and amphoteric constituents.

The color differentiation of the *cytoplasm* afforded by this polychromatic quality has proven extremely valuable, particularly for evaluating the regression grade and differentiating inflammatory changes.

The *nuclei* stain blue-gray, while *nucleoli* take on a dark blue and *cytoplasm* a greenish-blue (cyanophilic) color. Erythrocytes are bright red; in their mature state squamous cells (see p. 51) stain light red but sometimes also cyanophilic; and keratinized cells are, by contrast, an orange color.

1.12.1.1 Staining Technique According to Papanicolaou

a) **Spray fixation of smears:** Wash by immersing several times in 50% alcohol and in distilled water.

Staining with Harris hematoxylin (for nuclei)	6 min
Rinse with distilled water	1/2 min
Immerse in 0.25% HCl-alcohol	6 ×
Rinse in running water	6 min
Pass through increasing concentrations of alcohol (50%, 70%, 80%, 95%)	each 1/2 min
Staining with OG 6 (for cytoplasm)	1.5 min
2 × 95% alcohol (separate cuvettes!)	each 1/2 min
Staining with EA 50 (for cytoplasm)	1.5 min
3 × 95% alcohol (separate cuvettes!)	each 1/2 min
Absolute alcohol/xylene (in equal parts)	1/2 min
Xylene	1/2 min
Mount in Eukitt, Caedax or Canada balsam	

b) **Fixation of the smears with alcohol-ether** (see p. 13): Prior to being placed in the Harris hematoxylin solution, the smears are submersed in each of a descending series of alcohol concentrations:

80% alcohol	1/2 min
70% alcohol	1/2 min
50% alcohol	1/2 min
Distilled water	1/2 min

1.12.1.2 Compositions of the Stains for Papanicolaou's Method

Harris hematoxylin

Hematoxylin	1.0 g
Alcohol, 95%	10.0 ml
Aluminum or ammonium sulphate	20.0 g
Distilled water	200.0 ml
Yellow mercuric oxide	0.5 g
Glacial acetic acid	8.0 ml

OG 6 (Orange G 6)

Orange, G, 0.5% solution in 95% alcohol	100 ml
Phosphotungstic acid	0.015 g

EA 50 (Polychrome solution)

Light green S.F. yellowish	0.375 g
Bismarck brown	0.4 g
Eosin yellowish	2.5 g
Distilled water	50.0 g
Ethanol, 95%, pure	609.0 g
Methanol, pure	160.0 g
Phosphotungstic acid 1.7 g, dissolved in 95% ethanol	5.0 ml
Lithium carbonate solution, saturated	0.5 ml
Glacial acetic acid	1.0 ml

When the intensity of the staining weakens, new stains must be prepared. Although fresh staining solutions can be used to stain as many as 1000 smears (Soost 1978), they should be replaced every week regardless of the number of preparations stained.

For perfect staining, the following points must be observed:

- Filtration of Harris hematoxylin — 1 × daily
- Filtration of OG 6 and EA 50 solutions — every 2nd day
- The stains must be stored in a cool place, protected from light
- All alcohol solutions and rinsing media must be regularly topped up
- Only clear xylene may be used

If excessive staining of nuclei occurs (deep blue chromatin, fine structure no longer recognizable, blotchy appearance), the duration of staining with Harris hematoxylin must be reduced by 1–2 minutes and/or longer differentiation achieved by more frequent immersion (i.e. more than 6 × !) in 0.25% HCl-alcohol.

1.12.2 Staining According to May-Grünwald-Giemsa (MGG)

Only air-dried smears!	1 h
May-Grünwald solution: (methylene blue and eosin in methanol!)	4–5 min
Distilled water:	4–5 min
Giemsa solution: (azure-eosin-methylene blue solution)	20 min

1 ml Giemsa per 10 ml tap water

Air-dry

Xylene

Mount in Eukitt, Caedax or Canada balsam

If the staining procedure is to be successful, optimal acidity of the tap water is essential. Excessively acidic water results in a reddened picture, while alkaline water leads to the smears acquiring a blue tinge. The most exact way of achieving the correct acidity is to add two drops of Sörensen's reagent to diluted Giemsa solution (Soost 1978).

With MGG stain the nuclei appear blue-violet, the nucleoli dark blue, and the cytoplasm blue (cyanophilic) or pink (eosinophilic).

1.12.3 Hematoxylin-Eosin (H&E) Staining

Smears fixed by spray or alcohol-ether

Rinse in 50% alcohol	1/2 min
and in distilled water	1/2 min
Staining with Harris hematoxylin	3 min
Rinse in running water	10 min
Immerse in:	
70% alcohol	5 ×
1% HCl–70% alcohol	2–3 ×
70% alcohol	5 ×
70% alcohol (separate cuvettes!)	5 ×
3% ammonium/70% ethyl alcohol	2–3 ×
70% alcohol	5 ×
70% alcohol (separate cuvettes!)	5 ×
90% alcohol	5 ×
Staining with 1% eosin solution	1/2 min
95% alcohol	5 min
Immerse in:	
Absolute alcohol	5 ×
Xylene	2 1/2 min
Xylene (separate cuvettes!)	2 1/2 min

Mount in Eukitt, Caedax or Canada balsam

After H&E staining the nuclei appear pale blue, and the cytoplasm pink to rose-pink. The cell membranes always stand out clearly when the structure of a sheet of cells is orderly.

H&E stain is scarcely used in urological cytology, and even less frequently in prostatic cytology.

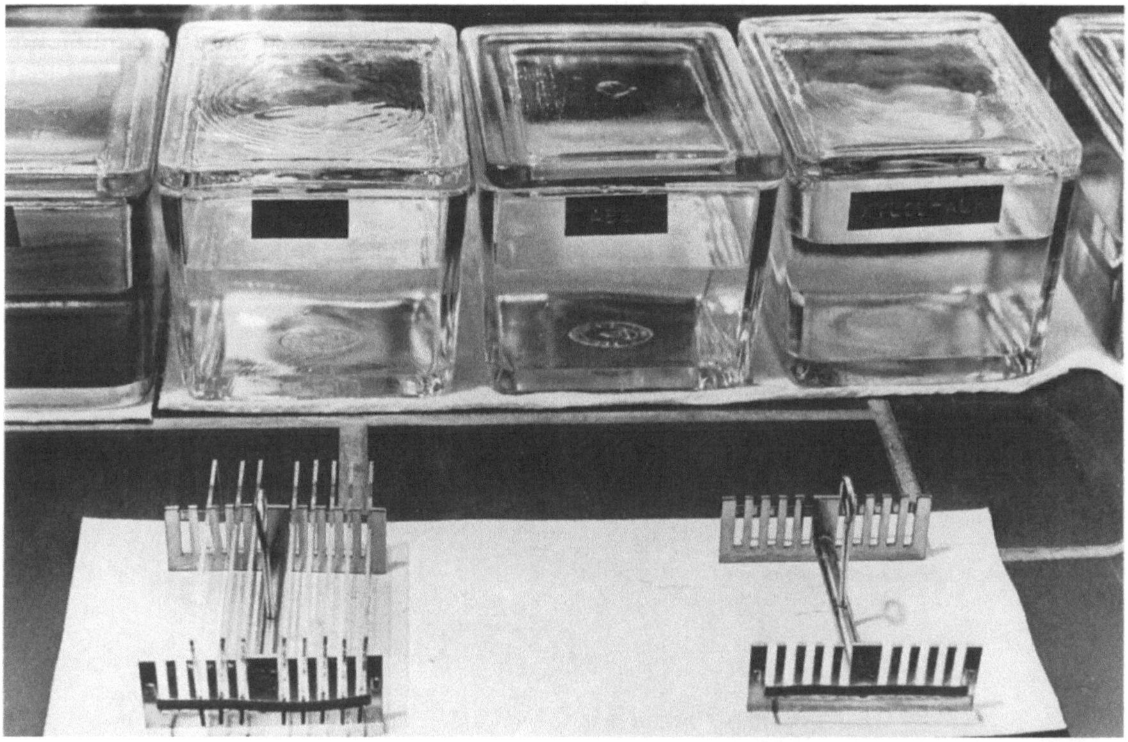

Fig. 19. Part of the stain bank, with two standard slide racks in the foreground

Fig. 20. Slide cabinet for clean and safe storage of cytological specimens

In order to make best possible use of the stain bank while carrying out staining on an independent basis, at least ten smears should be collected. They are placed in slide racks, which hold ten or more slides **(Fig. 19).**

Standard automated staining systems are of value only in large cytological laboratories which process a wide variety of materials.

The conventional staining solutions can be obtained ready for use from the manufacturer[1]. After diagnosis, the slides are stored, protected from dust, in slide cabinets[2] **(Fig. 20).**

1 Fa. Merck, Darmstadt; Cilag-Chemie, Abteilung Ortho-Alsbach

2 Fa. Technicon, Bad Vilbel

2 Cytological Microscopy

Since for the microscopic evaluation of prostate aspirates the technical equipment and certain specific details are as important as the aspiration technique itself and the preparation of the aspirate it is appropriate to describe them briefly.

2.1 Microscope (Fig. 21)

The parts of the microscope are:

- stand with object stage
- tube
- objective carrier
- optical system (eyepieces and objective lenses)
- illuminator

2.1.1 Stand with Object Stage

The stand (5) is the main component of the microscope. At the top of the stand (4) is the eyepiece tube (2) and just below it the revolving nosepiece with the objectives (3). The arm of the stand holds the condenser, which is adjustable and interchangeable (8) and the foot of the stand the lamp with collector, swing-out lens and field diaphragm (13).

The *object stage* (6) is attached below the objectives about halfway down the stand. It can be adjusted for vertical focussing of the image by means of coarse (14) and fine (1) focussing mechanisms.

The type of stage most commonly used for cytodiagnostics is the mechanical stage (6), with which the preparation can be moved along both the x- and y-axes.

2.1.2 Tube

The tube (2) contains the detachable eyepieces (1). Cytological diagnosis frequently requires considerable time and thus the *binocular tube* is to be preferred to the *monocular tube*, since it puts less of a strain on the eyes.

Discussion tubes have proven useful if a preparation is to be simultaneously evaluated by a second observer.

If a *vario tube* is introduced between the stand and the tube, the tube factor can be adjusted to any magnification between one and several times. An image focussed with the aid of an eyepiece with a graticule thus remains sharp within the whole zoom range, for example, in microphotography.

2.1.3 Objective Carrier

The objective carrier is attached to the top of the stand and is today produced only in the form of a so-called revolving nosepiece (3) with holders for 3 or more interchangeable objectives. The revolving nosepiece permits rapid exchange of objectives and has a "clickstop" locking mechanism.

2.1.4 Optical System (Eyepieces and Objectives)

2.1.4.1 Eyepieces

The eyepiece, which has the function of a magnifier, permits observation of the primary image furnished by the objective. It is important to distinguish between simple C eye-

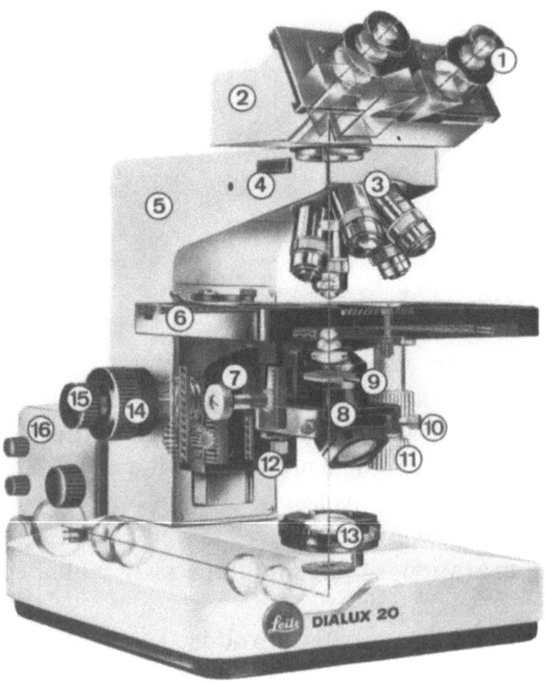

Fig. 21. Diagram of a laboratory microscope (Dialux 20, manufactured by Leitz, Wetzlar, West Germany). *1*, eyepieces; *2*, binocular tube; *3*, quintuple revolving nosepiece; *4*, slot for light filter; *5*, stand; *6*, mechanical stage; *7*, knurled knob for vertical adjustment of the condenser; *8*, condenser; *9*, lever for opening and closing the aperture diaphragm; *10*, knurled screw for centering the condenser; *11*, drive knobs for mechanical adjustment of the specimen; *12*, knurled screw for the upper stop of the condenser; *13*, knurled ring for setting the field diaphragm; *14*, coarse adjustment of the image; *15*, fine adjustment of the image; *16*, lamp housing (Determann and Lepusch 1980)

pieces and compensating plano eyepieces (Cpl). With Cpl eyepieces in combination with a plano-achromatic objective it is possible to achieve better color correction and correction of curvature. These properties make them particularly suitable for cytodiagnostics.

The most common eyepiece magnifications are:

8 ×
10 × and
12.5 ×

Widefield eyepieces with an extended angle of view increase the field of view and are far less tiring on the eyes, particularly during extended use of the microscope.

Highpoint eyepieces have a high pupil position for spectacle wearers, enabling them to obtain optimal image quality without removing their spectacles.

2.1.4.2 Objectives

The objectives are classified according to their degree of color correction and curvature of field. There is a distinction between simple achromats, fluorite systems and apochromats and plano objectives (plano achromats and plano apochromats), with which the visual field is uniformly sharp from the centre to the edge. These plano achromats (magnification: 40 ×) are very expensive but afford the maximum sharpness of detail.

The best rendering of a preparation as regards color is achieved with an apochromat, which brings 3 colors of the spectrum to a common focus. The color correction afforded by the different types of objective is engraved in the objectives, likewise the abbreviation "Pl" for plano objective (e.g. Pl Apo).

The *magnifying power, aperture, tube length* and *coverglass thickness* of an objective, i.e. its most important properties, are engraved in its mount.

The engraving
Pl Apo 40/0.65; 170/0.17
thus gives the following information:
objective with a flat field of view, 40-fold magnification and a numerical aperture of 0.65

The figure 170 gives the length of the tube in mm and the figure 0.17 the prescribed coverglass thickness in mm.

Conventional objective magnifications are:

2.5 ×	40 ×
10 ×	100 ×
25 ×	

The combination of objective and eyepiece gives the total magnification of the microscope. It depends on the numerical aperture, which should be between 500 and 1,000 times that engraved on the objective.

The numerical aperture is the measure of the optical resolving power and thus of the performance of an objective.

The range between 500 and 1,000 times the aperture is termed the "useful total magnification". It can be calculated as follows for objectives with the specifications 40/0.65: the useful total magnification is 325–650 times (500 × 0.65 to 1,000 × 0.65) the aperture. Thus with an eyepiece affording 12.5-fold magnification the total magnification attainable is 500 times greater, which is optimal for the aperture of the objective.

On the other hand an eyepiece with 25-fold magnification would give a total magnification 1,000 times greater, which would far exceed the "useful total magnification". Only limited evaluation of a cytological preparation would be possible because the image would not be sharp.

Objectives with a high resolving power have a wide aperture (over 0.95) and require the use of *immersion oil* if their resolving power is to be fully utilized. This substantially reduces the refraction of the rays emerging from the coverglass and the total reflection from the surface of the coverglass, or even prevents it completely. Thus if immersion oil is employed the angle of incidence of rays falling on the objective is far greater.

The maximum numerical aperture with which immersion oil can be used is about 1.40.

Immersion oil must not be allowed to resinify and must comply with certain requirements as regards refractive index and dispersion.

2.1.5 Illuminators

Microscope illuminating units comprise the following parts:

- low voltage illuminator with a centrable lamp, collector and swing-out lens (16);
- field diaphragm;
- condensor with aperture diaphragm (8, 9).

The light source may be in the form of detachable lamps, built-in illuminators or illuminators with their own stands. Detachable illuminators are usually attached to the foot of the microscope stand.

The illuminating system and field diaphragm of a *built-in illuminator* are integrated into the foot of the stand while the lamp is either in the stand or attached to the outside in the form of a detachable lamp housing (16). These illuminators are fitted with 50–100 Watt halogen lamps for normal microscopy and photomicrography.

The lamp housing has slots into which the filters are mounted. A heat filter must be inserted between the collector and the filter slots.

2.2 Brightfield Microscopy

Brightfield microscopy is most suitable for prostatic cytology. However correct usage of the microscope is vital if important details are to be detected – in prostatic aspirates this applies especially to the nucleus. Correct usage of the illuminating system and accurate adjustment of the aperture and field diaphragms and condenser are also of decisive importance.

2.2.1 Aperture Diaphragm

The aperture diaphragm is part of the condenser. It is used to stop down the image and the light source and to regulate resolution, contrast and depth of focus of the image, but may *not* be employed to regulate brightness. This must be done via a transformer.

In prostatic cytology images always contain both *high* and *low contrast* structures. The aperture diaphragm should therefore be set as follows:

open wide to start with, then gradually stop down to roughly 2/3 of the aperture, to allow satisfactory evaluation of the less well-differentiated structures. Further stopping-down increases the contrast, but at the same time the image becomes darker.

2.2.2 Field Diaphragm

The field diaphragm is used to alter the cross-section of the ray bundle in the object plane. The illuminated field in the object can be narrowed so that it coincides with the field of view of the microscope. It should therefore be opened until it just disappears out of the field of view.

2.2.3 Condenser

Before the microscope is used the condenser should always be adjusted to its optimal setting. First turn in the top of the condenser. Then move it to its topmost position by turning the knurled knob used to adjust the condenser vertically (7), close the field stop with the appropriate knob (13) and focus the image in the field of view using the vertical adjustment device. Center the field diaphragm with the aid of the knurled screws used to center the condenser (10), i.e. move it into the middle of the field of view. The field and aperture diapragms are thus opened.

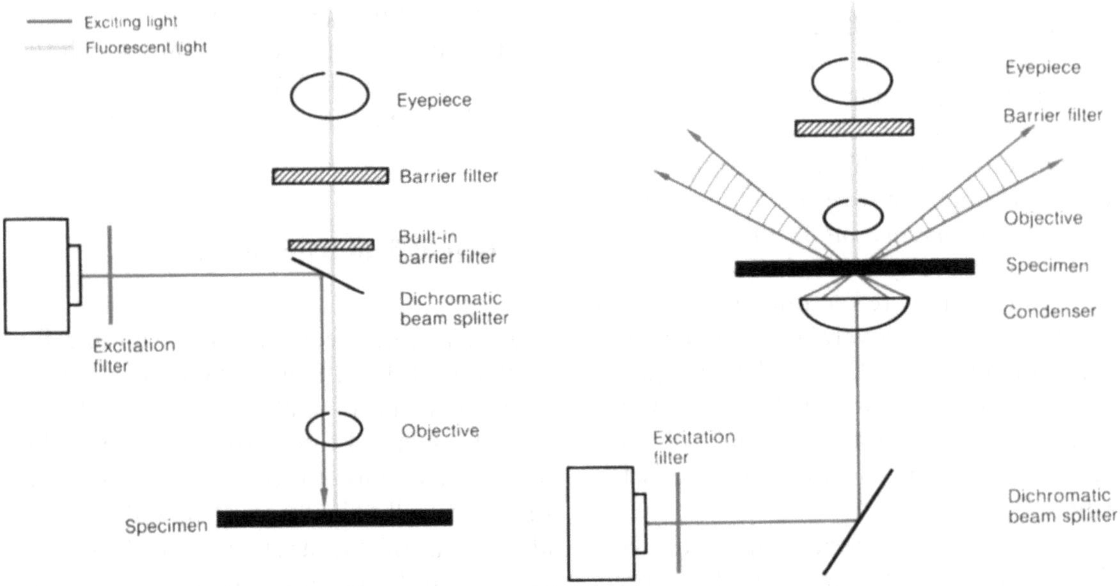

Fig. 22. The principles of incident light fluorescence (*left*) and transmitted light fluorescence (*right*) (KOCH 1972)

2.3 Fluorescence Microscopy

In addition to brightfield microscopy fluorescence microscopy – especially in combination with DNA cytofluorometry – is also of considerable importance as a method. Two systems are suitable, namely

- incident light fluorescence **(Fig. 22)**
- transmitted light fluorescence **(Fig. 22)**.

2.3.1 Principle of Fluorescence Microscopy

The *exciting light* is produced by a light source (high-pressure xenon lamp, 75 or 150 watt or ultra-high pressure mercury discharge lamp, 50 or 100 watt). Filters which can be permeated by the exciting light from the light source only *(excitation* or *primary filters)* are fitted into the lamp housing.

2.3.1.1 Incident Light Fluorescence (Fig. 22)

In incident light fluorescence the excitation rays are reflected through the microscope objective onto the preparation by means of dichromatic beam splitters, thus permitting utilization of the full aperture of the objective for excitation. The intensity of the excitation radiation depends on the numerical aperture of the objective employed, which is why microscope objectives with a high numerical aperture are the most suitable.

Special incident light illuminators are fitted with revolving holders for the excitation filters so that the filters can be rapidly exchanged.

2.3.1.2 Transmitted Light Fluorescence (Fig. 22)

In transmitted light fluorescence the excitation rays pass through a condenser (darkfield or brightfield) before reaching the preparation.

In order to allow observation of the fluorescence emitted by the fluorochromated preparations a barrier filter adjusted to the wavelength of the fluorescent light (secondary filter) is placed in the optical path. This absorbs or reflects the excitation radiation and gives a dark background.

If optimal utilization of the excitation energy is to be achieved the light source and condenser must be accurately centered.

2.4 Guidelines for Brightfield Microscopy

The magnifications of the cytological illustrations in this book are equal to those of microscopic images obtained by combining the following *eyepieces* and *objectives:*

objective:	10 ×, 40 ×, 63 × and 100 × (oil immersion)
eyepiece:	10 ×

The scales of the illustrations can be calculated by multiplying the magnifications of the objective and eyepiece and correspondingly range from 100 to 1,000-fold magnification. *Widefield eyepieces* with a magnification of 10 magnitudes have proved to be the most suitable.

A good command of microscopy with its various possibilities is vital for the evaluation of prostatic aspirates. Only if this requirement is met can one make a reliable diagnosis of carcinoma of the prostate with primary malignancy grading, determine and classify therapy-induced signs of regression, and differentiate between the various forms of prostatitis and atypias and clearly defined carcinoma. If command of the techniques is inadequate, cytoplasm, nuclear size and the accentuation or even prominence of nucleoli cannot be reliably evaluated.

A final diagnosis should only be made at a total magnification of 400 magnitudes (eyepiece: 10 ×, objective: 40 ×).

2.5 Procedure

First examine each smear of an aspirate with an *objective with 10-fold magnification* to establish whether the cellular material and sheets of epithelial cells suffice, since fewer than 20 sheets of cells from one prostatic lobe are *not enough* for a reliable cytological diagnosis (see p. 11)!

If this preliminary examination shows the smear to contain enough sheets of cells, observe through an objective with a magnification of 25 magnitudes, moving it in a "meandering" motion, starting at the edge of the coverslip. A magnification of 250 magnitudes is usually high enough for atypias to be easily recognizable. However, they must then without fail be thoroughly inspected for details, particularly the density and structure of the chromatin, the presence of nuclear membranes and the size and shape of the nucleoli, using a 40-fold objective and simultaneously increasing the illumination. Only when this has been done can a final diagnosis be made.

Finally, use of a more powerful objective, e.g. one with a magnification of 63 magnitudes (= a total magnification of 630 magnitudes) permits optimal evaluation of the nucleolar size, nuclear polymorphism and chromatin density.

If the sheets of cells do not appear sharp this is usually due to dirt on the eyepiece, objectives or slide.

The eyepieces should be cleaned with a fine, soft linen cloth and the objectives with a cotton bud soaked in benzol. Xylene has proven suitable if the lenses are badly soiled.

The *slides* should be cleaned with alcohol only.

3 Normal Findings

3.1 Individual Cells, Sheets of Cells and Background

Normal epithelial cells of the prostate occur in the aspirate in sheets of varying size. These sheets are so loosened from the walls of the acini or from the excretory ducts by the forced injection of air at the start of the aspiration, that they can be drawn into the needle by means of the vacuum created in the syringe.

It is characteristic of a *normal sheet of prostatic epithelium* that even preliminary screening of the smear at low magnification (100- to 125-fold) reveals clear definition of the cells and a regular sheet structure **(Fig. 23).**

While the dissociation of individual cells at the periphery of the sheets seldom occurs, *fissures* or *circular spaces* are seen more frequently. Although this may create an impression of glandular ducts, they are, in fact, circumscribed areas torn open by the suction applied during aspiration **(Figs. 23, 24).**

The *background of a smear* containing normal sheets of epithelium is predominantly clear. Isolated epithelial cells and secretion are scarcely encountered. *Erythrocytes* are found frequently in a technically satisfactory aspirate, but they do *not* impede diagnosis. In fact, correct diagnosis is usually possible even in the presence of large amounts of blood, if the aspirate otherwise contains sufficient sheets of epithelium (**Fig. 40,** p. 54).

Normal epithelial cells display the following characteristics:

- *clearly recognizable cell borders*
- *small, round, uniform nuclei*
- *regular positioning of the nuclei*
- *loose, granular chromatin structure*
- *small, scarcely recognizable nucleoli*
- *no shifts in the nuclear-cytoplasmic ratio*

The *honeycomb pattern* of normal prostatic epithelium is very typical owing to the well defined cell borders, but nonetheless not an obligatory characteristic. *In addition, normal sheets of epithelium are typified by clear, in part finely granulated cytoplasm and by regularity of the nuclear-cytoplasmic ratio* **(Figs. 24–26).**

Even in normal prostatic epithelium, however, the *honeycomb pattern* is not always uniform throughout an entire sheet of cells, and occasionally only a portion will be structured in this way **(Fig. 27).** Moreover, an intact honeycomb pattern is *in itself no proof that the epithelium is normal,* since the pattern can be preserved in well differentiated (G I) and occasionally even in less well differentiated (G II) carcinomas **(Figs. 70, 76, 77, 96).**

Consequently it is not the findings at the cellular level or the structure of the sheets of cells that are decisive for classification as normal epithelium, but rather the cytological parameters relating to the nucleus.

3.2 Nuclei

In relation to the volumes of the cells, the nuclei are so small that there would be room for at least two in a normal prostatic epithelial cell.

Classification of the nuclei into

small,
medium, and
large

is only possible qualitatively, their relation to the cytoplasm representing a suitable aid in this assessment.

Nuclei that appear very small in comparison to the cytoplasm, and in addition very compact owing to a homogeneous chromatin structure, are typical of *atrophic epithelium.* In sheets of this nature, cells that have an optically empty cytoplasm, i.e. *without* a nucleus, are also frequently found **(Fig. 25).**

Elongated nuclei in a normal sheet of prostatic epithelium are a rarity, and a consequence of smearing.

When *MGG staining* is used, normal nuclei appear distinctly *larger* than with Papanicolaou staining **(Fig. 30a vs 28).**

The *arrangement of the nuclei* is regular. They mostly lie centrally in the cytoplasm, only occasionally overlapping locally. The *nuclear membrane* is generally recognizable as a soft contour and displays no infolding **(Fig. 29).**

3.2.1 Chromatin

The chromatin is loosely granular, and distributed evenly throughout the nucleus. It is for this reason that the nuclei are transparent after Papanicolaou staining, whereas they tend to appear homogeneous when MGG staining is employed **(Figs. 27, 29a vs 30a).**

In atrophic epithelium, by contrast, the chromatin is homogeneous and so compact **(Fig. 25)** that the intense blue coloration gives rise to an impression of overstaining of the nuclei.

3.2.2 Nucleoli

In the nuclei of normal prostatic epithelial cells, nucleoli are only very weakly distinguishable at a magnification of 400- to 500-fold, while at still higher magnification the occasional one is slightly 'conspicuous'. If nucleoli can be recognized at all, only one is present per nucleus – and its relation to the nucleus is *never* such that it appears 'conspicuous' **(Fig. 29b).**

As with the nuclei, *determination of nucleolar size* is merely qualitative. Confident classification of the nucleoli as

scarcely distinguishable
small
conspicuous
prominent

requires much experience; **comparison with the diameter of the cell membranes within the same group of cells has proven to be of help in this.** If, using this criterion, the diameter of the nucleoli at 400- to 500-fold magnification is approximately the same as that of the membranes, they are to be classified as 'conspicuous' and hence as only slightly atypical **(Fig. 35).** If, however, at this magnification the nucleoli show a greater diameter than the cell membranes, they are to be adjudged 'prominent' **(Fig. 37a).**

When just a very occasional 'prominent' nucleolus is seen in a group of epithelial cells in which 50 or 60 nuclei contain 'small' or 'conspicuous' nucleoli, this is a basis for diagnosing *atypia* only, *not* prostatic carcinoma **(Figs. 37a, 38b).**

Decisive for the classification of prostatic epithelium as 'normal' is the proof that all the typical parameters used to assess normality do indeed pertain in the majority of the nuclei.

It is hardly ever possible to distinguish a well differentiated prostatic carcinoma (G I) from normal prostatic epithelium solely by means of the magnification (100- to 125-fold) used for screening. Consequently random samples of all sheets of cells whose structure does not arouse suspicion at this magnification must be examined at a higher magnification too (400- to 500-fold!).

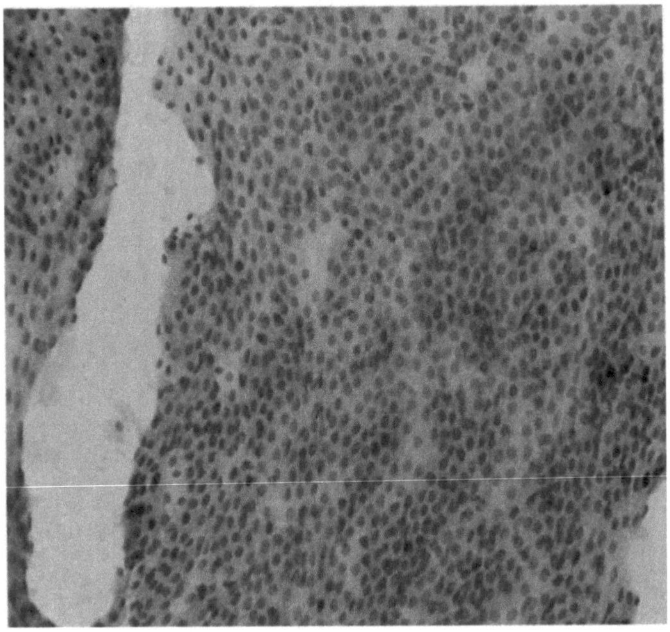

Fig. 23. Normal prostatic epithelium at the low magnification used for preliminary screening. Uniform nuclei with a regular, orderly arrangement; clear definition of the sheet, with fissure formation. ×100

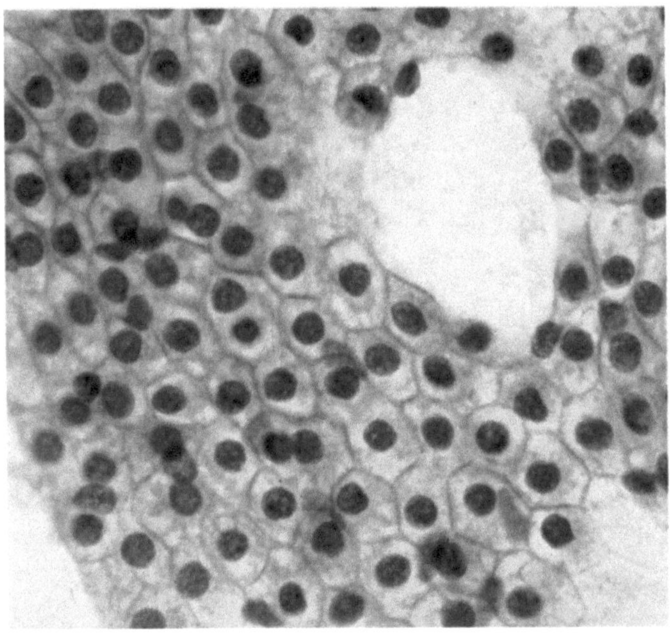

Fig. 24. Normal sheet of cells displaying the typical honeycomb pattern and fissure formation. ×400

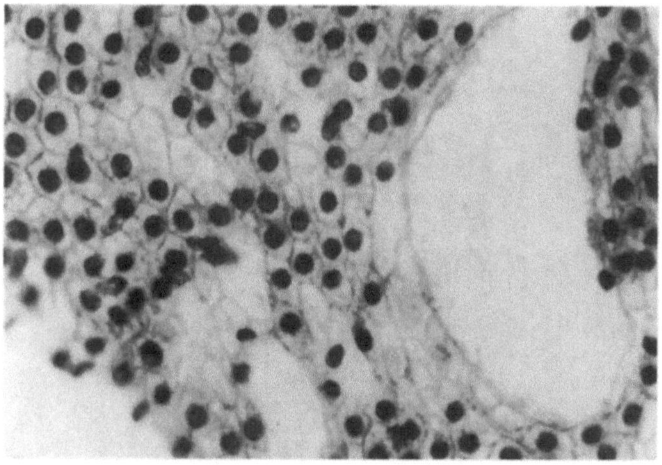

Fig. 25. Findings in atrophy of the prostate: relatively small nuclei, most of which appear dark. Several anuclear cells, clearly recognizable honeycomb pattern. ×100

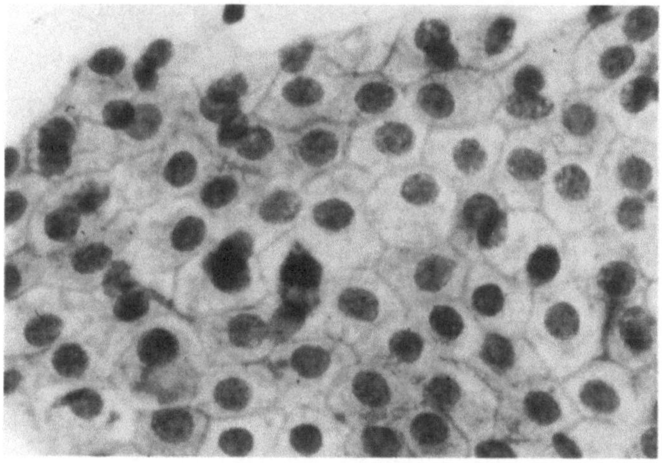

Fig. 26. Normal sheet of cells with a slightly disturbed arrangement. Uniform nuclei, chromatin loosely structured, scarcely distinguishable nucleoli, honeycomb pattern. ×400

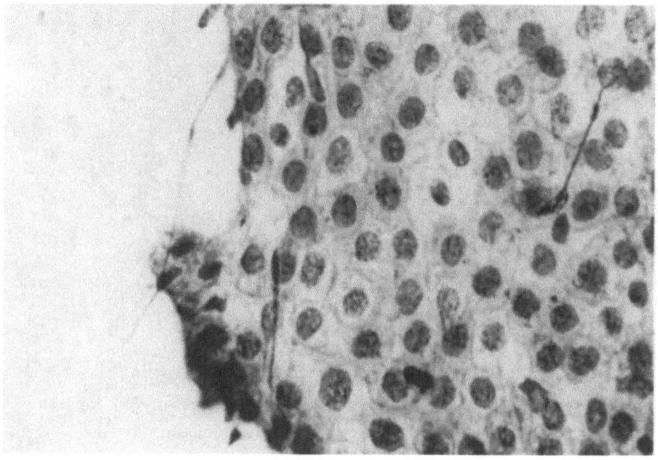

Fig. 27. Normal sheet of epithelium with typical nuclear parameters and isolated thread-like nuclear extensions. Honeycomb pattern suggested only locally. ×400

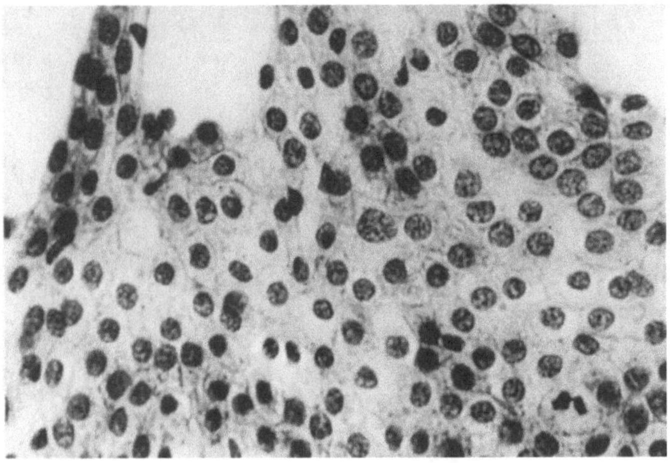

Fig. 28. Normal sheet of cells with distinct nuclear membranes and many atrophic, dark nuclei. Cell borders absent in places. ×400

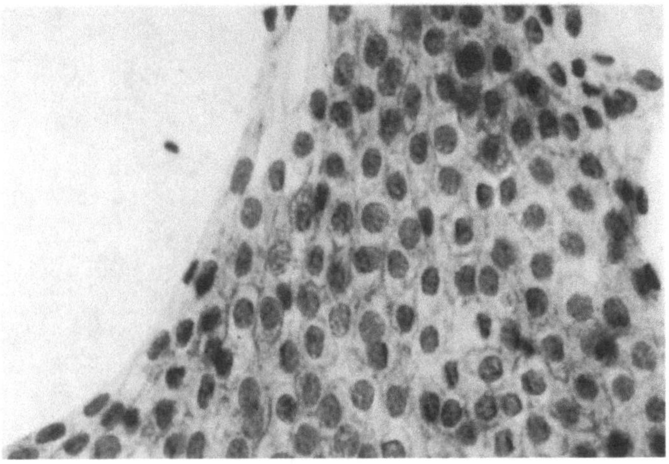

Fig. 29a. Normal sheet of epithelium showing a disorderly arrangement in the upper part of the picture and isolated atrophic nuclei. ×400

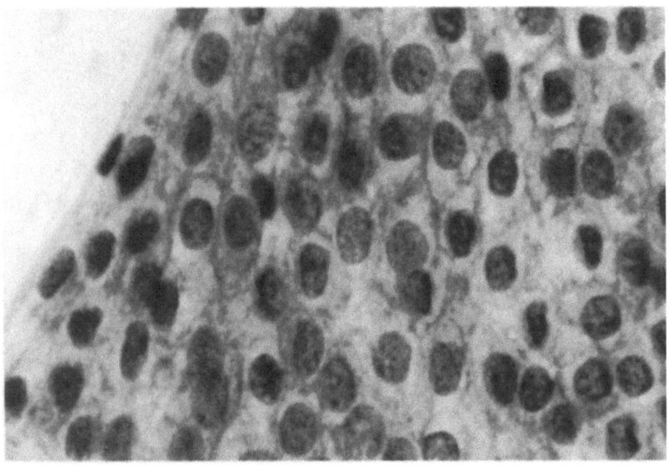

Fig. 29b. Same sheet at higher magnification. Apart from in the atrophic nuclei, a loosely granular chromatin structure; scarcely distinguishable nucleoli. ×630

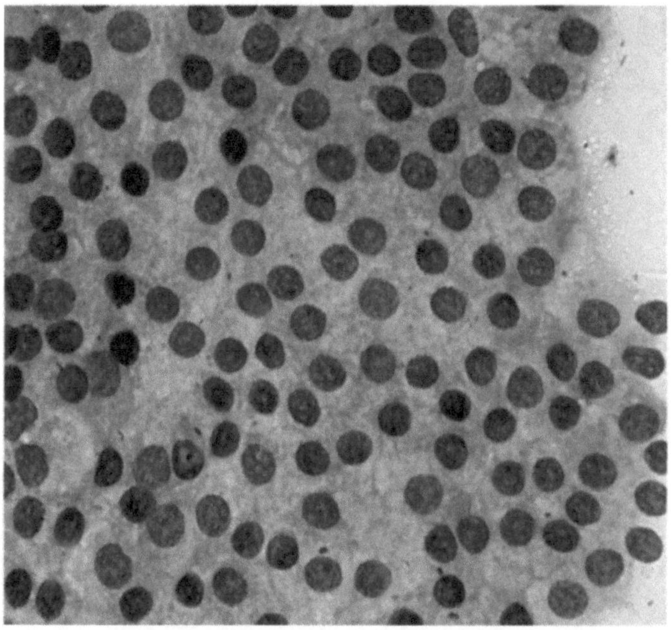

Fig. 30a. Normal sheet of cells, stained according to May-Grünwald-Giemsa. ×400

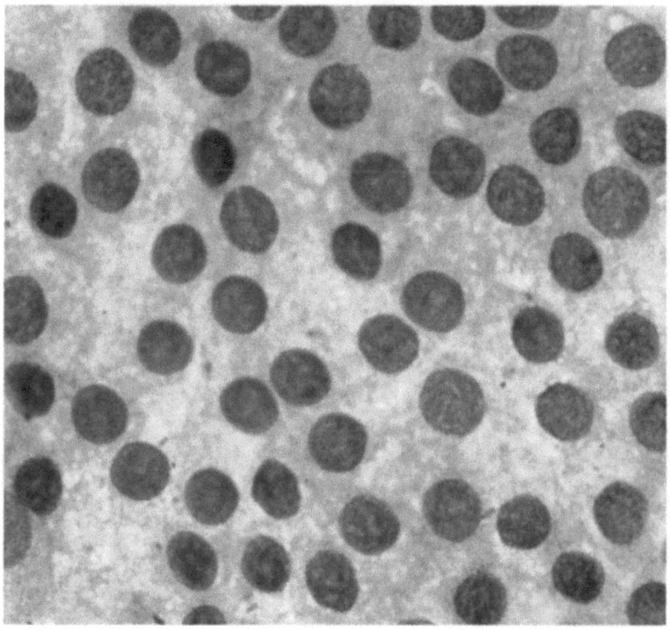

Fig. 30b. Same sheet at higher magnification. ×630

4 Atypia

Deviations from normal cellular and nuclear structure can be demonstrated in the prostate by cytological investigation of the epithelium, just as they can in other organs. Since the spectrum of such deviations ranges from findings that can still just be regarded as normal to very pronounced atypia, various classifications based on general cytological criteria have been introduced (Editorial, Acta Cytologica 1964).

The most frequently used system of classification is that according to Papanicolaou (1954). This system makes it possible to classify with precision findings that are normal, atypical, suggestive or cytologically conclusive of carcinoma by assigning them to five groups or classes **(Table 5).**

Table 5. Cytological classification (PAPANICOLAOU 1954)

I	Absence of atypical or abnormal cells
II	Atypia, but without evidence of malignancy
III	Suggestive of, but not conclusive for, malignancy
IV	Strongly suggestive of malignancy
V	Conclusive signs of malignancy

Although Papanicolaou's classification was originally intended for gynecological cytology, it can be adopted as it stands for the classification of prostatic aspirates.

In correspondence with this system, findings may be classified into groups I–V ('Pap I–V'), the designation 'Pap V' to indicate definite prostatic carcinoma today having been replaced by the diagnosis 'prostatic carcinoma' in conjunction with the cytological grade of differentiation (G I–G III) (see Chap. 7).

4.1 Classification According to Papanicolaou

4.1.1 Papanicolaou I (Normal Findings)

The designation 'Pap I' corresponds to a cytologically normal aspirate without any appreciable degree of atypia.

4.1.2 Atypia (Papanicolaou II–IV (Table 6)

The different forms of atypia are allocated to groups II–IV in accordance with the extent to which deviations – especially of the nuclei – from the norm are present in relation to the following parameters:

nuclear size and shape
anisokaryosis
nuclear arrangement
chromatin
nuclear membrane
nucleolus

4.1.2.1 Papanicolaou II (Atypia Without Evidence of Malignancy) (Figs. 31–34)

The prostatic epithelial cells show only minor deviations from normal findings, which

Table 6. Cytological parameters for atypia of the prostate

	Nuclear size/shape	Aniso-karyosis	Nuclear arrangement	Chromatin	Nuclear membrane	Nucleoli
Pap II	Small, round	Slight	Slightly disturbed	Loosely granular	Intact	Slightly conspicuous
Pap III	Small, round to oval	Moderate	Slightly disturbed	Condensed	Some intact	Conspicuous
Pap IV	Enlarged, some round, some oval	Pronounced, diffuse	Markedly disturbed	Markedly condensed, homogeneous	Few intact	Conspicuous, some prominent

above all occur with respect to variation in *nuclear size* and *arrangement*. The *chromatin* is in part still loosely granular, in part condensed; most of the cellular and nuclear membranes are still intact **(Fig. 31).**

The nucleoli are conspicuous only locally and are not prominent even at higher magnification! The cytoplasm is clear **(Figs. 33b, 34).**

4.1.2.2 Papanicolaou III (Suggestive of, but not Conclusive for, Malignancy) (Figs. 35, 36)

The forms of atypia to be assigned to this group can be of a most diverse nature, so that reliable classification is occasionally difficult. Findings of this type correspond to medium-grade atypia, which may occasionally be pronounced within individual groups of cells.

The subclassification of these forms of atypia into three classes (III a–III c), proposed by Papanicolaou for gynecological cytology, is unnecessary in prostatic cytology.

It is characteristic of this group of atypias that there is more or less pronounced

- disturbance of nuclear arrangement
- variability in nuclear size
- and, above all, considerable changes within the nucleus itself **(Figs. 35, 36)** (see nuclear parameters, **Table 6)**

The nuclei are still small, but not all of them are round; rather some are round to oval. The *chromatin* is frequently condensed, but the *nuclear membranes* are often still clearly recognizable **(Fig. 36).**

Even at 400-fold magnification, the *nucleoli* are distinctly conspicuous, though not yet 'prominent'. Occasionally a nucleus contains two nucleoli **(Figs. 35a, 36a).**

The above-mentioned forms of atypia are also very frequently encountered in certain forms of prostatitis (see Chap. 12), subsequent to iatrogenic trauma of the prostate (surgery, biopsy) and after local infarctions.

If *no* indications of inflammation accompany these forms of atypia, control investigations should be performed at regular 6-monthly intervals, since it is not yet clear whether a carcinoma cannot in fact later develop from such alterations in prostatic epithelium.

4.1.2.3 Papanicolaou IV (Strongly Suggestive of Malignancy) (Figs. 37–39)

Classification into this category depends upon still more striking changes in the nuclei.

Nuclear arrangement and *structure* are only partially still regular, and in total are significantly disturbed. The *chromatin* is unequivocally homogeneously dense, the nuclei vary considerably in size and shape (anisokaryosis) and the *nuclear membranes* are often no longer distinguishable **(Figs. 37, 38).**

The nucleoli are *conspicuous, and locally already prominent. Isolated nucleoli may have lost their circularity.* Binucleolate nuclei are now seen more frequently **(Figs. 37b, 39).**

It is not rare for such findings also to be encountered in cases of *granulomatous nonspecific* and of *specific (tuberculous) prostatitis* (see pp. 168, 169), whereas they do not occur in other inflammatory disorders of the prostate.

If the smear in question yields no indication of prostatitis, the findings must be considered strongly suggestive of prostatic carcinoma but are not proof thereof. Consequently a control biopsy after 4 weeks is imperative.

The most reliable criterion for differentiating between so-called Pap IV characteristics and prostatic carcinoma is that in the former the changes do not occur throughout a group of cells but rather only locally.

4.2 Atypical Hyperplasia

In the event of Pap IV findings, *differential diagnosis* must take particular account of *'atypical hyerplasia'* (MOSTOFI and PRICE 1973) – a *'marked atypia'* (MILLER and SELJELID 1971) which histological and autoradiographic investigations (KASTENDIECK 1980; HELPAP 1980) have shown to be closely related to prostatic carcinoma and which should probably be regarded as precancerous or as early carcinoma.

Histologically, moderately enlarged and polymorphic nuclei with chromatin of varying density are described, in addition to irregular glandular architecture (KASTENDIECK 1980).

GAETINI and TRENTINI (1978) diagnosed cytologically 'atypical hyperplasia' in 31 of 141 cases of prostatic hyperplasia; these diagnoses were confirmed histologically. They named the following cytological parameters as diagnostically relevant:

- a clean, light background to the smear
- an acinar-like cell arrangement
- indistinct cell borders
- scanty and granular cytoplasm
- negative cytoplasmic fluorescence after acridine orange staining
- uniform, round or oval nuclei
- nuclear hyperchromatism
- dense and finely granular chromatin structure
- small, often unrecognizable nucleoli
- no mitosing cells

These nuclear alterations correspond to cytological Pap IV findings.

In four of five personal cases of histologically confirmed 'primary atypical hyperplasia' we were able to demonstrate such changes cytologically.

According to the findings to date, patients with 'atypical hyperplasia' are to be regarded as at risk for carcinoma and should therefore undergo regular control by aspiration biopsy at intervals of 6 months.

If, after one or two further aspiration biopsies, the same Pap IV type atypias persist, a precautionary punch biopsy should be performed.

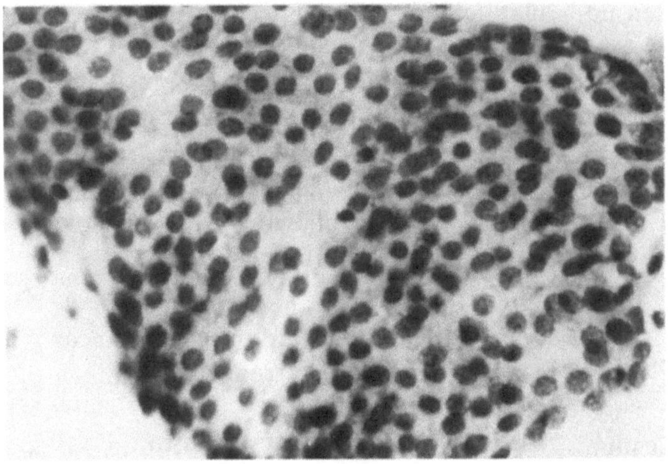

Fig. 31 a. Sheet of prostatic epithelial cells with Pap II atypia: low-grade variation in nuclear size and disturbance of nuclear arrangement. ×100

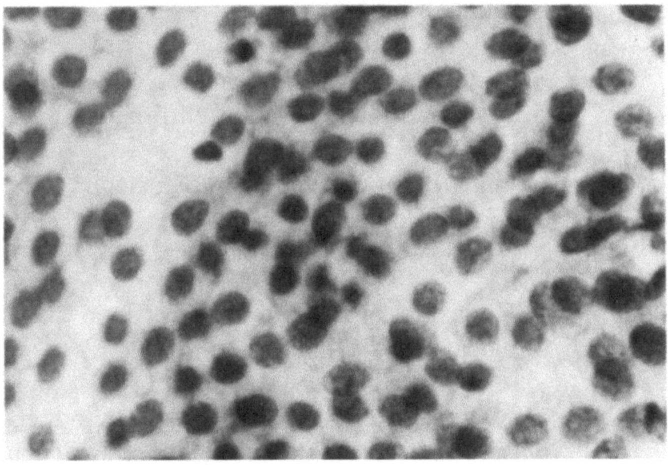

Fig. 31 b. Same sheet at higher magnification: no enlarged nucleoli. Nuclear chromatin predominantly still loosely granular. ×400

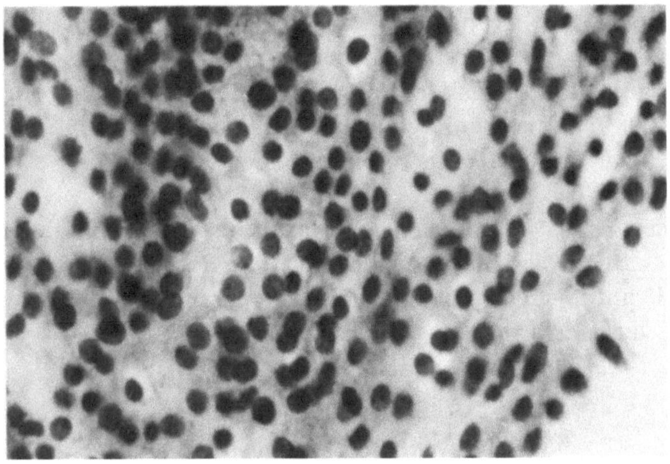

Fig. 32 a. Pap II atypia alongside numerous atrophic nuclei. ×100

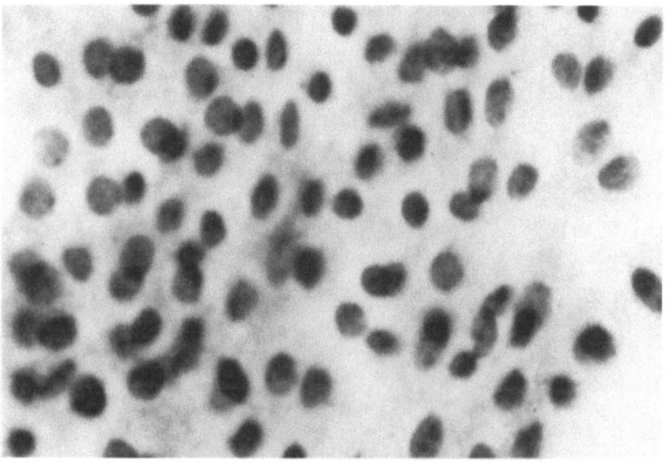

Fig. 32b. Same sheet at higher magnification. ×400

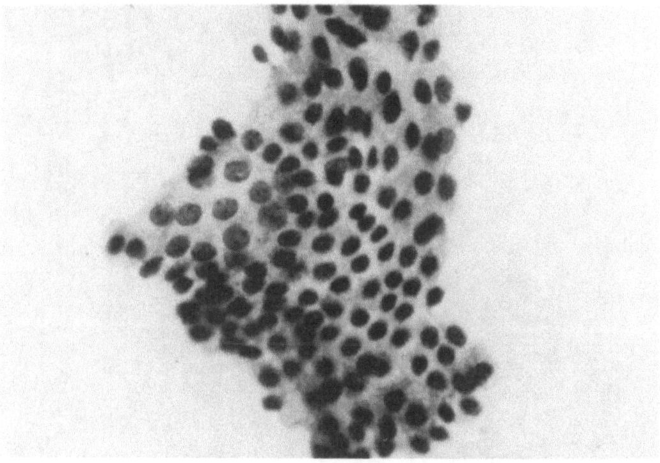

Fig. 33a. Pap II atypia with locally conspicuous nucleoli. ×100

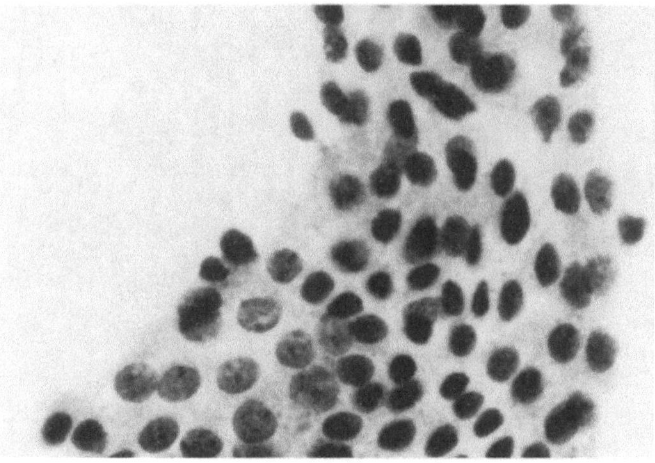

Fig. 33b. Same sheet at higher magnification. Clear representation of the nuclei containing conspicuous nucleoli at the lower left of the picture. ×400

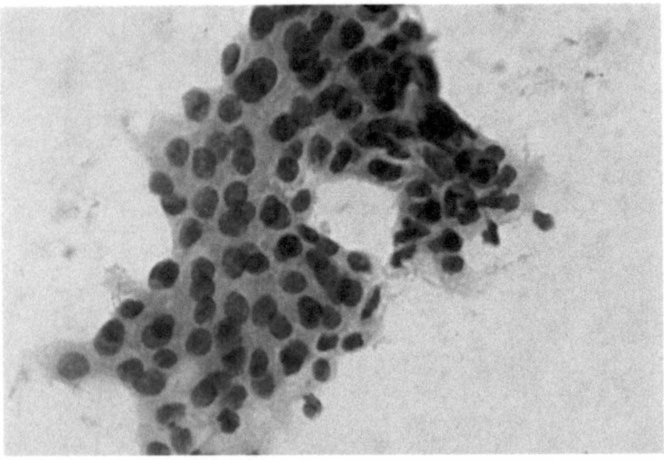

Fig. 34a. Pap II atypia. ×400

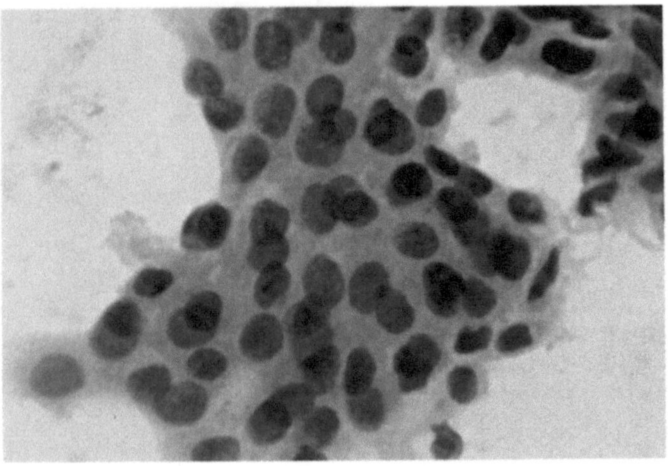

Fig. 34b. Same case at higher magnification. ×630

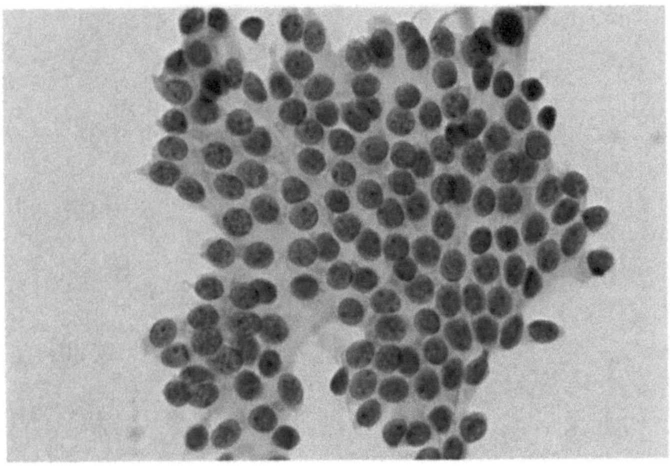

Fig. 35a. Pap III atypia: disturbed nuclear arrangement, moderately condensed chromatin and conspicuous nucleoli. ×400

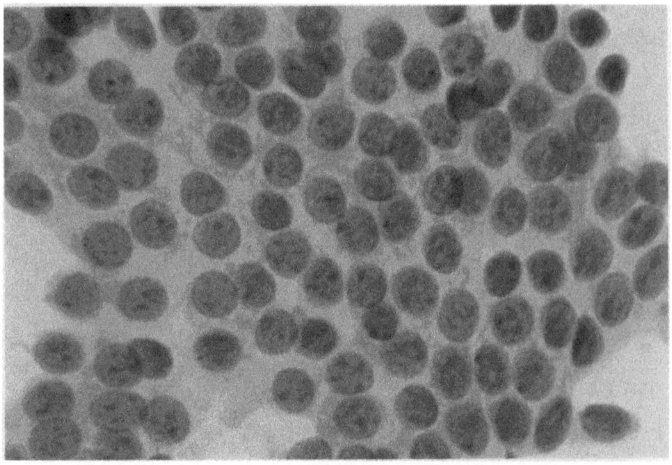

Fig. 35b. Same case at higher magnification: isolated fine chromatin clumping is now evident, together with low-grade loss of circularity of many nucleoli. ×630

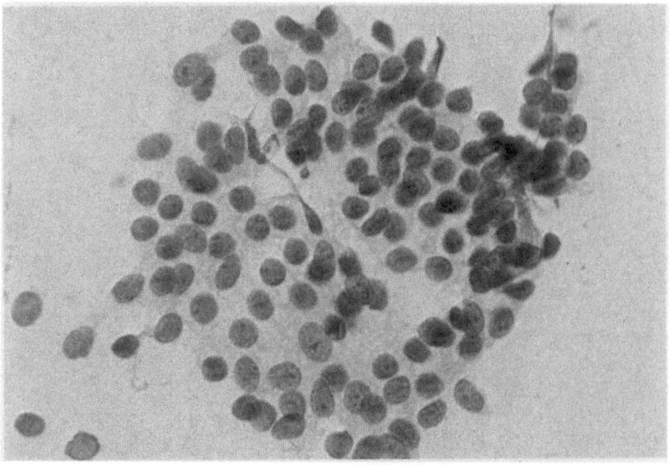

Fig. 36a. Marked variation in nuclear size and disturbed nuclear arrangement. Conspicuous nucleoli are seen throughout the group. ×400

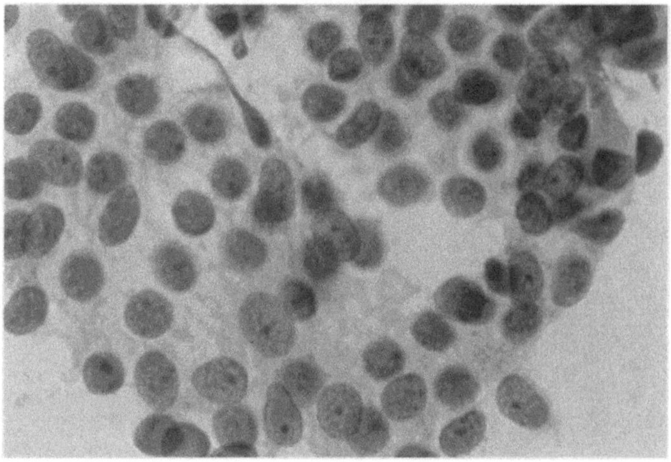

Fig. 36b. Same case at higher magnification: nucleoli are still only conspicuous, not prominent. The nuclear chromatin is only slightly condensed. Many nuclear membranes are still clearly distinguishable. ×630

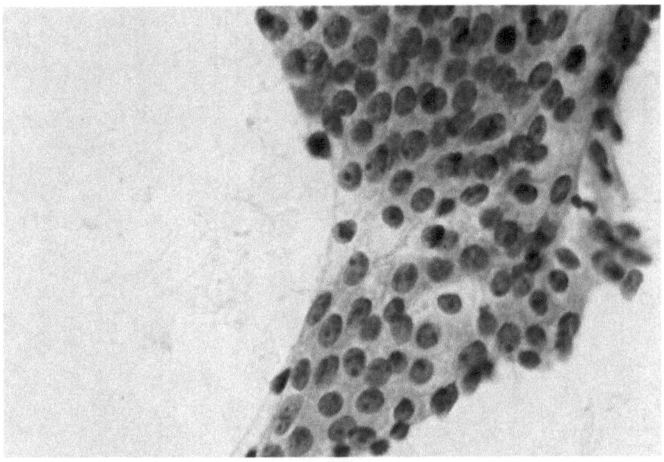

Fig. 37a. Pap IV atypia: more severe disturbance of nuclear arrangement in the upper part of the picture. Marked variation in nuclear size; conspicuous, and in part prominent, nucleoli. ×400

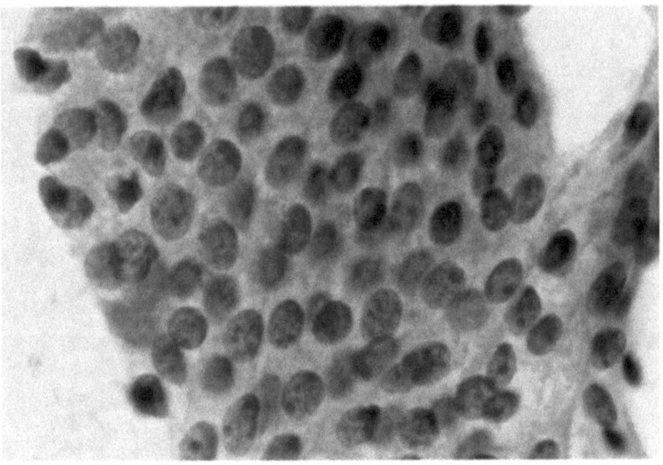

Fig. 37b. Same case at higher magnification: the loss of the uniform nuclear pattern is more easily recognizable; in the lower part of the picture there are several nuclei that contain prominent nucleoli; disturbance of the chromatin distribution is seen in places. ×630

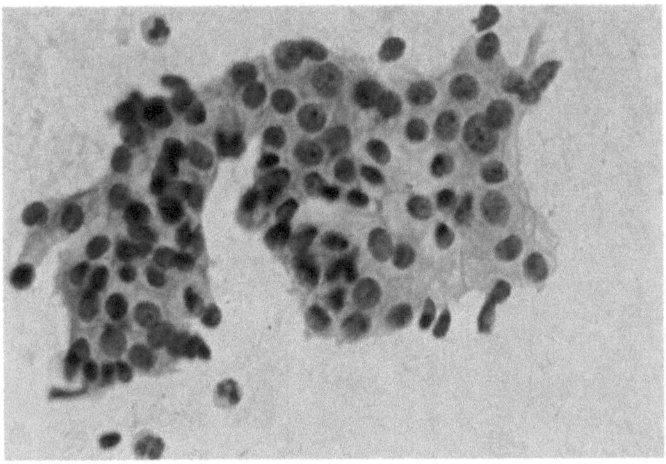

Fig. 38a. Pap IV atypia: marked variation in nuclear size, moderately condensed chromatin and more pronounced disturbance of the nuclear arrangement; conspicuous nucleoli, in places beginning to become prominent. ×400

Fig. 38b. Same group of cells at higher magnification: the condensed chromatin and the local prominence of the nucleoli are now more clearly recognizable. ×630

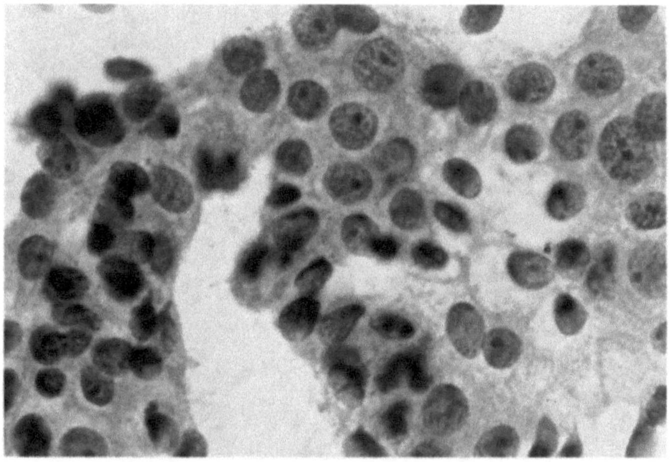

Fig. 39a. Pap IV atypia: marked variation in nuclear size and increased chromatin density; conspicuous to prominent nucleoli. ×400

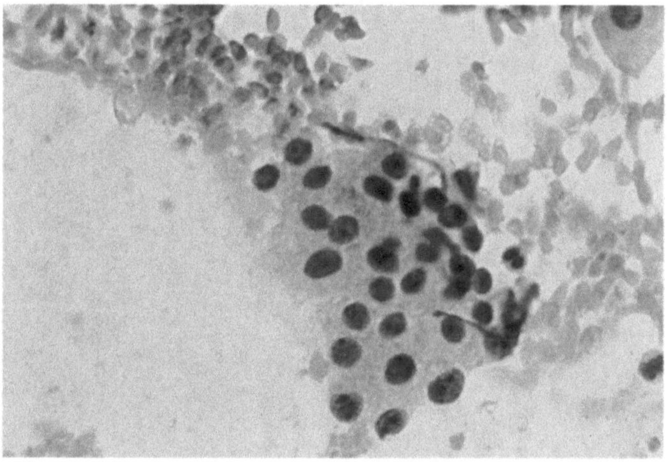

Fig. 39b. Same group of cells at higher magnification: some nucleoli display loss of circularity; in parts there is clumping of the chromatin; some nuclei are round to oval in shape. Differential diagnosis: atypical hyperplasia or grade I carcinoma. ×630

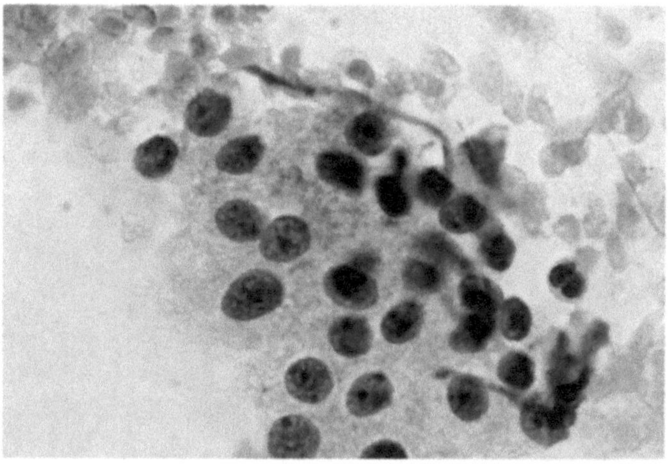

5 Secondary Findings

In addition to prostatic epithelial cells, a variety of other cells are found in smears of prostatic aspirates. These include:

- erythrocytes
- seminal vesicle epithelial cells
- epithelial cells of rectal mucosa
- urothelial cells
- metaplastic squamous epithelial cells
- histiocytes

5.1 Erythrocytes
(Figs. 39, 40)

Moderate quantities of blood in an aspirate are usually located at the periphery of groups of epithelial cells or surround them. The erythrocytes are often faded and then appear greenish yellow **(Fig. 39)** or bright red **(Fig. 40)** after staining according to Papanicolaou.

The assessment of groups of epithelial cells encircled by erythrocytes is not impaired when only a moderate quantity of blood is present. If, however, predominantly blood is aspirated, exact diagnosis is scarcely possible even in the presence of quantitatively sufficient cellular material. This is because the erythrocytes generally overlie the epithelium to such an extent that fixation of the latter is very difficult, with the result that it is no longer possible to recognize details with precision subsequent to staining. Smears containing a large amount of blood should therefore be discarded at the time of aspiration.

5.2 Seminal Vesicle Epithelial Cells
(Figs. 42–46)

Cells derived from seminal vesicle epithelium, which tend to be aspirated especially when the suspect area lies in the basal region of the prostate, must be distinguished from prostatic epithelial cells with great care owing to the danger of false-positive diagnosis of carcinoma. Typically, seminal vesicle cells are found in well preserved sheets, with occasional dissociation of individual cells from the periphery of the sheet **(Figs. 42–44).** Their *nuclei* are in part round, in part markedly oval or triangular, and always larger than those of normal prostatic epithelium **(Fig. 41).** In addition, more or less pronounced nuclear polymorphism is regularly found **(Figs. 43, 44).**

The *arrangement of the nuclei* is irregular, *the nuclear-cytoplasmic ratio* is shifted in favor of the nucleus, and the *cell borders* occasionally display a honeycomb pattern **(Fig. 45).**

The *chromatin structure* is predominantly dense. The *nuclear membranes* are generally still clearly visible, even though they are delicately formed **(Fig. 44).**

Although the cytological characteristics of seminal vesicle cells described above are also found in severely atypical or even carcinomatous prostatic epithelium, seminal vesicle cells can be classified with certainty on the basis of the following characteristics, which are especially pronounced when Papanicolaou staining is used:

- intracytoplasmic yellowish-brown pigment granules and vacuoles **(Figs. 42–45)**
- relatively small nucleoli **(Figs. 41–45)**
- distinct nuclear membranes **(Figs. 43–45)**
- polygonal shape of some nuclei **(Figs. 42b–44)**

Differential diagnosis is also facilitated by the presence of mucus or isolated spermatozoa in the smear.

5.3 Epithelial Cells of Rectal Mucosa
(Figs. 47–52)

The aspiration of epithelial cells from the rectal mucosa is usually caused by a technical error, namely withdrawing the Franzén needle from the prostate *without* prior equalization of pressure, i.e. with the plunger still drawn back. Rectal mucosal cells can, however, also be aspirated directly, above all when the needle is not pushed far enough into the prostate; this occurs not infrequently when either the prostate or the suspect area itself is small.

Admixtures of stool are rare, and only to be expected when the ampulla recti is filled with excrement **(Fig. 46)**.

With some experience it is possible to recognize abundantly aspirated sheets of rectal mucosal cells *macroscopically* within fresh smears, in that the smear is traversed by fine, short gray-white strands.

Cytologically, definite recognition of rectal mucosal cells is an absolute prerequisite for the reliable evaluation of prostatic aspirates, since in the primary diagnosis of carcinoma they can, depending upon the 'plane of section' of the aspiration, easily be mistaken for prostatic epithelial cells!

The appearance of rectal mucosal cells in smears can vary greatly in accordance with the level being depicted. Although their basic structure is similar to that of prostatic epithelial cells, they differ from the latter by virtue of (among other things) their local balloon-like vacuolation, which resembles the cytoplasm of tangentially punctured goblet cells **(Figs. 47, 51)**. Confusion with cytoplasmic vacuoles is possible, but can almost always be avoided if one searches for the *classical criteria* of nuclei of rectal mucosal cells after staining according to Papanicolaou:

- round to oval nuclear shape **(Figs. 47–49, 52)**
- thick, clearly distinguished nuclear membranes, which are occasionally indented (coffee-bean nuclear shape) **(Figs. 47b, 49)**
- loose chromatin structure **(Figs. 47b, 49, 52)**

It is easiest to recognize sheets of rectal mucosal cells when whole glandular lumina are met transversely. This results in the typical rosette appearance with the classical basally situated nuclei in the goblet cells **(Fig. 50)**.

5.4 Urothelial Cells
(Figs. 53, 54)

Urothelial cells are found in the prostatic aspirate when the urinary bladder or the prostatic urethra is directly punctured. Owing to the multilayered nature of the urothelium, diverse forms of urothelial cells are seen; they stem either from the surface (superficial cells) or from deep layers (intermediate and parabasal cells).

The urothelial cells that are most frequently present in aspirates are those from the deep layer. Their classification is unproblematic: occurring in predominantly small groups, the cells are striking on account of their slender cytoplasm, which often terminates in a tail-like extension **(Fig. 53a)**. *The cytoplasm* is basophilic and finely granular; fine pigment granules are sometimes seen **(Fig. 53b)**.

The *nuclei* are either located in the center of the cytoplasm or are displaced rather towards its periphery. They are oval, uniform and have an easily recognizable membrane.

Nucleoli can often be distinguished, and not infrequently they are slightly conspicuous. Occasional binucleolated nuclei are to be seen.

The *nuclear-cytoplasm ratio is regular* **(Fig. 53b)** (see, by contrast, the section 'Urothelial Carcinoma', p. 153).

Urothelial cells from the upper layer are larger and in part polygonal. The *cytoplasm* is well developed, and the tail-like extension is rarer **(Fig. 54)**. In cytoplasm which is paler perinuclearly, vacuoles are typical, as are *binucleation* and/or *multinucleation*. Consequently in extreme cases the picture is one of pronouncedly *multinucleated giant cells*.

5.5 Metaplastic Squamous Epithelial Cells
(Fig. 55)

'Metaplasia' may be understood as the replacement of one type of epithelium by another through the transformation of certain cells. This metaplasia may be benign, for instance when the cylindrical glandular epithelium of certain organs undergoes transformation into normal squamous epithelium (squamous cell metaplasia). Metaplastic epithelial cells develop from the basally situated reserve cells of epithelium and display varying degrees of maturation. Metaplasia differs unambiguously from malignancies by virtue of the uniformity of the nuclei in size and shape; in addition the chromatin is loosened (TAKAHASHI 1981).

In the differential diagnosis it is easy to distinguish between the aforementioned *benign metaplasia* and metaplasia of malignant epithelial cells, e.g. carcinoma cells of the urinary bladder epithelium versus squamous carcinoma cells, which are frequently demonstrated during increasing dedifferentiation of these tumors and which carry a poor prognosis.

Untreated prostatic carcinomas show no tendency towards squamous cell metaplasia, but primary squamous cell carcinoma of the prostate is to be expected in less than 1% of cases – 0.21% according to DHOM (1980; it is thus an extremely rare form of carcinoma.

Cytologically demonstrable squamous cells in groups of epithelial cells from a prostatic carcinoma always originate from normal prostatic epithelial cells or adenoma cells, as do the squamous cells found in the smears of prostatic carcinomas treated in various ways (see Chap. 9).

It is assumed that it is true squamous cells that are involved in benign squamous cell metaplasia in the prostate, and transformation of such cells into carcinoma cells has never been reported to date (KASTENDIECK and ALTENÄHR 1975).

Animal experiments indicate that the squamous cells evidently develop not from the prostatic epithelial cells active in secretion but from the basal reserve cells (which are regarded as ambivalent), and above all from the dorsolateral portions of the prostatic lobes lying near to the urethra (HOHBACH 1977).

The following are considered to be the causes of benign squamous cell metaplasia:

- *hormonal influences* (estrogens, progestogens, anti-androgens, bilateral orchiectomy)
- *special forms of treatment for carcinoma* (Estracyt®)

- *prostatic infarction*
- *transurethral resection*

Cytologically, the nuclei of squamous cells are always larger than those of prostatic epithelial cells. They are round to oval, with a loosely granular chromatin structure and scarcely any 'conspicuous' nucleoli. The nuclear membrane is delicate **(Fig. 55).**

The *cytoplasm* is cyanophilic or eosinophilic, and occasionally appears yellow or orange due to a high content of keratin or its precursors (SPIELER et al. 1976). It is partly finely homogeneous and partly finely granular. Especially in the perinuclear area it appears empty and 'clear as water' owing to the dissolving out of glycogen ('glycogenic cells') brought about by the preparation of the smear.

5.6 Sheets of Keratin
(Fig. 56)

Benign metaplastic squamous epithelial cells should not be mistaken for sheets of keratin, which stem from the peri-anal area of the rectum, stain yellowish-orange, and are anucleate **(Fig. 56).**

5.7 Histiocytic Giant Cells
(Figs. 57–59)

The often considerable size of the histiocytes and their wealth of nuclei may mimic groups of epithelial cells from a prostatic carcinoma, especially during treatment. Isolated histiocytic giant cells in the prostate are invariably an expression of a strong resorptive reaction to inflammatory disorders or surgical procedures on the prostate.

5.7.1 Cytoplasm

The cytoplasm of histiocytic giant cells is very abundant and the cell borders are occasionally difficult to distinguish **(Figs. 57–59).** Foreign matter in the cytoplasm, such as nuclear debris from prostatic epithelial or inflammatory cells, droplets of secretion, and granules, results in a foamy granular picture (so-called foam cells), especially in the various forms of prostatitis (see Chap. 12).

5.7.2 Nucleus

The multinucleation of the giant cells is striking. They can contain between 10 and 100 nuclei. Not infrequently the formation of Langhans' giant cells occurs **(Fig. 57).**

If a giant cell contains a particularly large number of nuclei **(Fig. 58),** differentiation from prostatic epithelium may be difficult at low magnification.

The nuclear structure is decisive in the identification of multinucleated histiocytic giant cells:

- The *nuclei* are oval and lie eccentrically in the cytoplasm **(Figs. 57, 58).**
- The *chromatin* is very loose and therefore appears extremely 'pale' **(Fig. 58);** only occasionally are small clumps of chromatin present.
- The *nuclear membrane* is always clearly defined **(Figs. 57, 58).**
- The *nucleoli* are generally conspicuous, and in individual nuclei even slightly prominent **(Fig. 58).**

Accordingly, atypia and malignancy can be distinguished on the basis of the histiocytes' typical nuclear chromatin structure and the very distinct nuclear membrane.

5.8 Intracytoplasmic Granules

The cytoplasm of prostatic epithelial cells contains granules of varying stainability as well as lipoid material. The precise composition of these granules has not yet been established. They are sometimes plentiful, though only in normal epithelium, and are turned a brownish color by Papanicolaou's stain **(Fig. 67).** With MGG stain they appear blue-violet, while they are a bright blue after staining with alcian blue (STAEHLER et al. 1975).

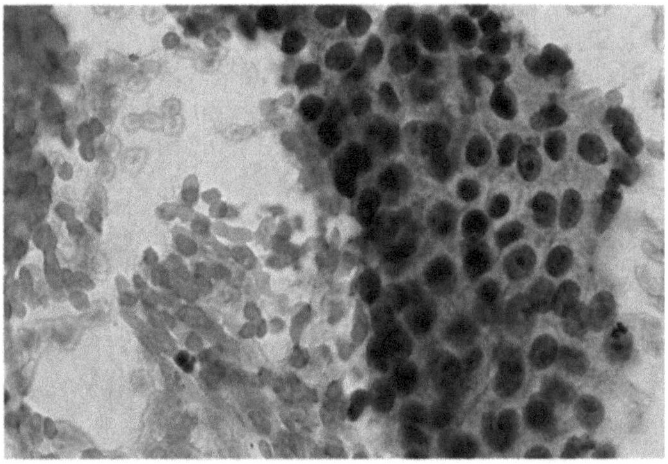

Fig. 40. Numerous erythrocytes, some of which are faded, alongside a sheet of cells from a grade I carcinoma. ×400

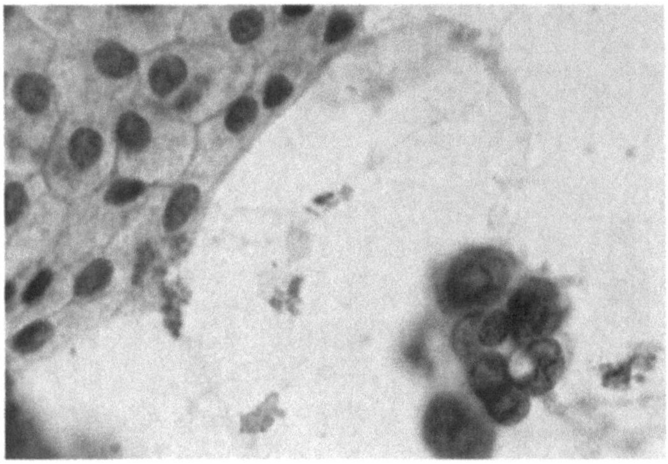

Fig. 41. Group of seminal vesicle cells on the right of the picture; compare with normal prostatic epithelium on the left. Despite their polymorphism, the nuclei of the seminal vesicle cells do not display prominent nucleoli. ×630

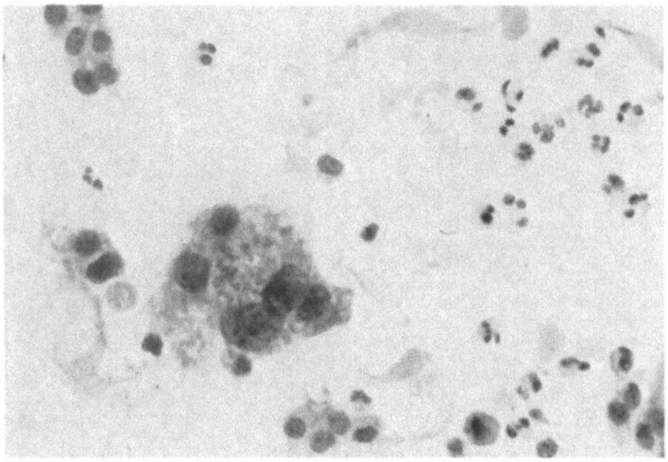

Fig. 42a. Small sheet of seminal vesicle cells with numerous pigment granules in the cytoplasm. *Right,* leukocytes; *upper left,* normal prostatic epithelium. ×400

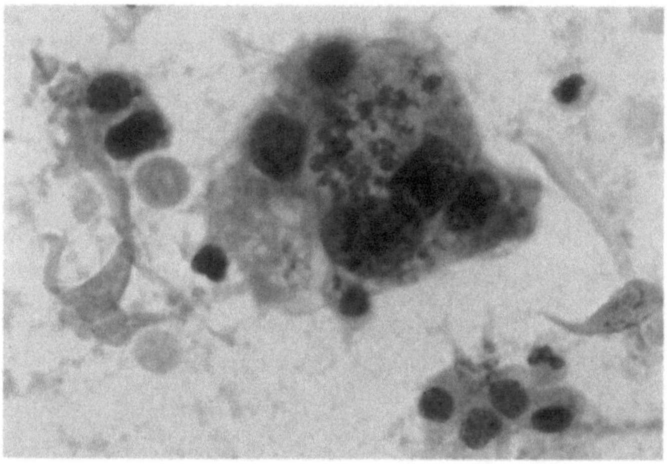

Fig. 42b. Same case at higher magnification. Despite considerable nuclear polymorphism, the nucleoli in the seminal vesicle cells are still only conspicuous, not prominent. ×630

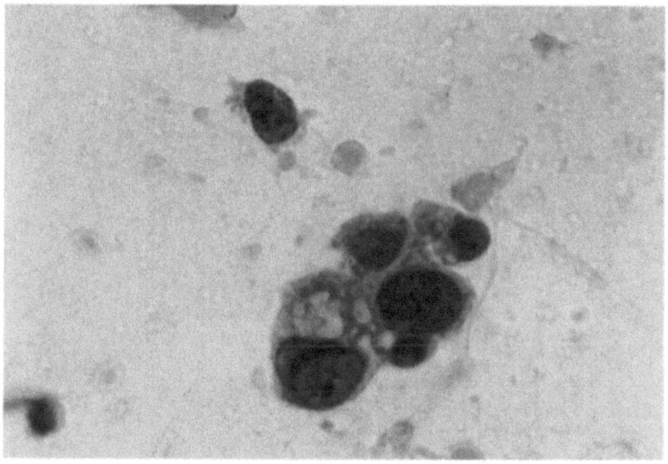

Fig. 43. Small sheet of seminal vesicle cells with typical criteria and distinct nuclear membranes. Nuclear shape in part polygonal. ×630

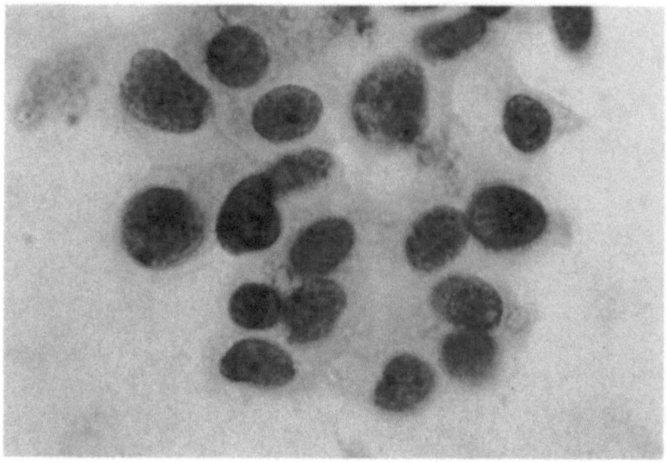

Fig. 44. Large sheet of seminal vesicle cells at higher magnification: considerable chromatin density is clearly visible. Nevertheless, the nucleoli are not prominent, and are conspicuous only locally. Most nuclear membranes are easily recognizable. Oil immersion ×1000

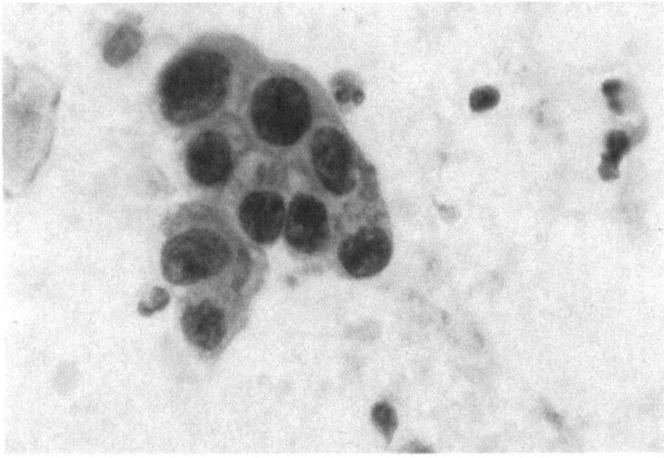

Fig. 45. Sheet of seminal vesicle cells with only moderate nuclear polymorphism and a 'paving stone' appearance of the nuclei. No prominent nucleoli. Typical pigment granules. ×630

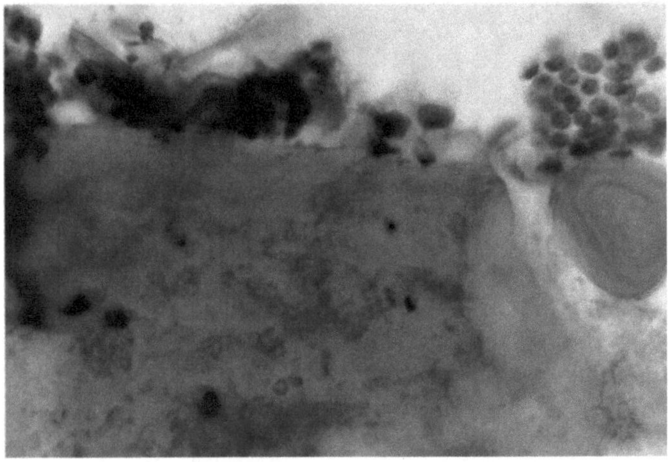

Fig. 46. Aspirate with admixture of stool. Small sheet of prostatic epithelium at the upper right of the picture. ×400

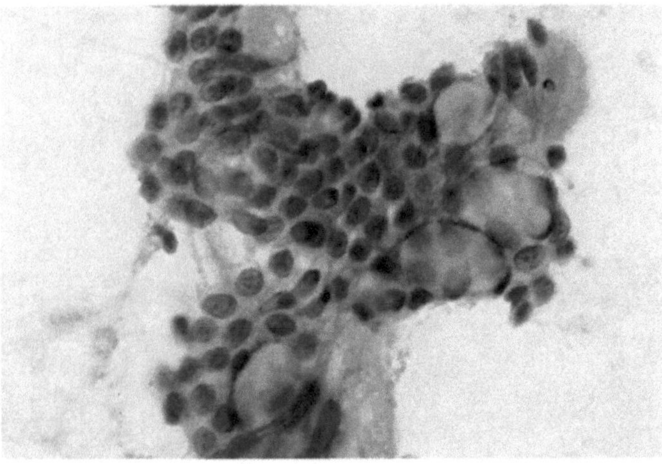

Fig. 47a. Tangentially smeared sheet of cells from the rectal mucosa. Local balloon-like vacuolation of the cytoplasm of the goblet cells. Small, round to oval nuclei with clearly defined nuclear membrane. ×400

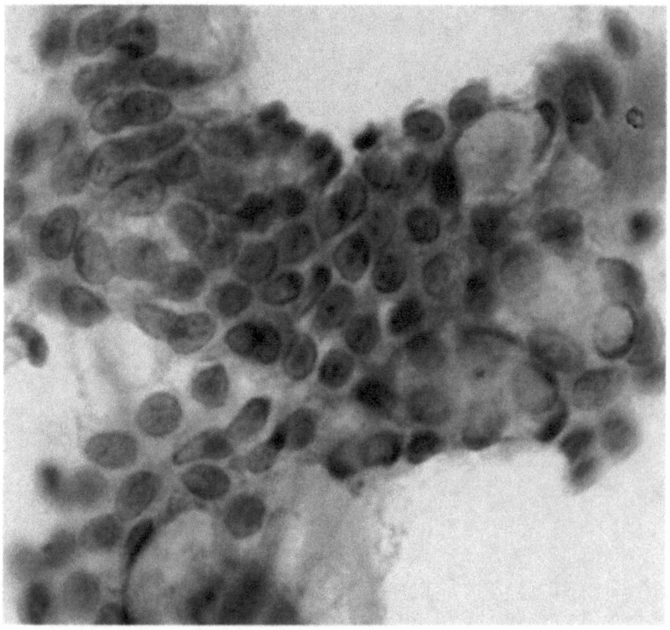

Fig. 47b. Same sheet at higher magnification: the exceedingly thick nuclear membranes are clearly seen. Loose chromatin structure; typical coffee-bean shape of the nuclei in places. ×630

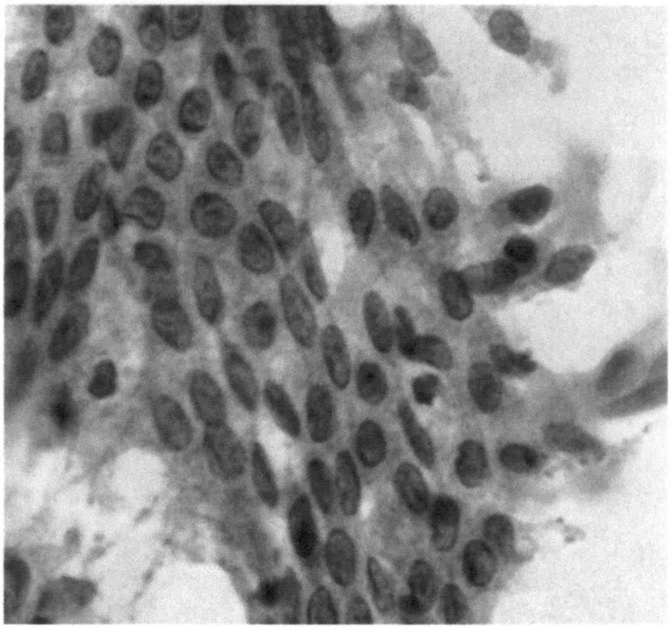

Fig. 48. Sheet of rectal mucosal cells without the typical cytoplasmic vacuolation but with the classical nuclear characteristics. ×630

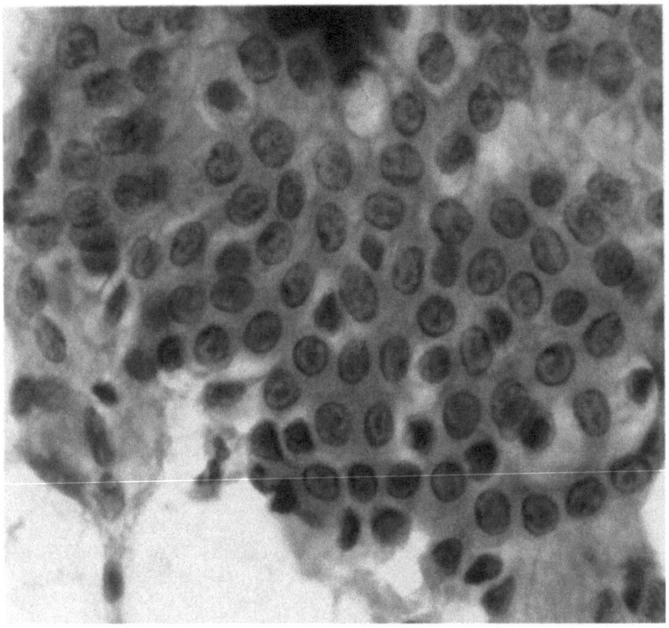

Fig. 49. Sheet of rectal mucosal cells with numerous typical indentations of the thick nuclear membranes (coffee-bean shape!) and a loose chromatin structure. ×630

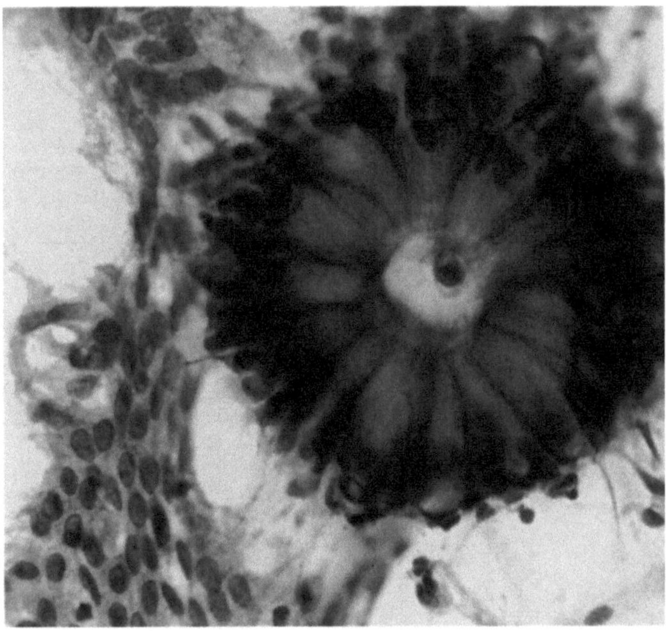

Fig. 50. Large sheet of rectal mucosal cells with a transected glandular duct displaying the typical rosette appearance. Nuclei basally situated; cytoplasm of the goblet cells elongated and extensive. ×630

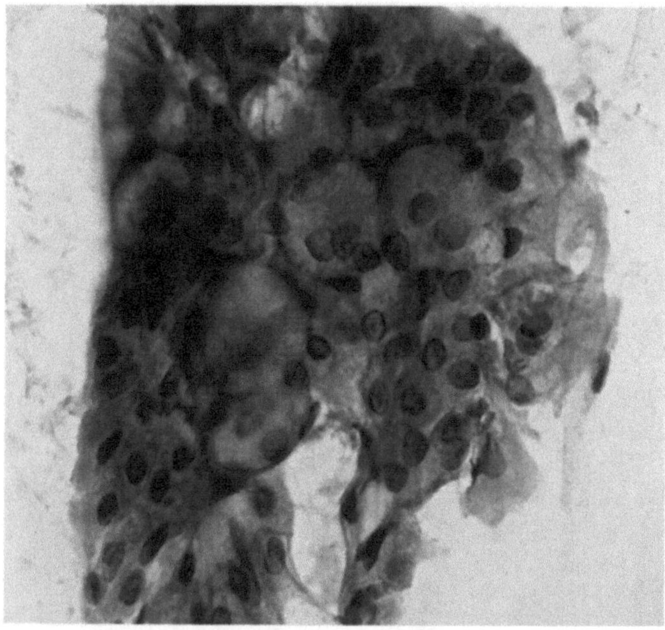

Fig. 51. Group of rectal mucosal cells. Despite the pronounced overlapping of the nuclei, the cells are identifiable unequivocally on the basis of the typical features. ×400

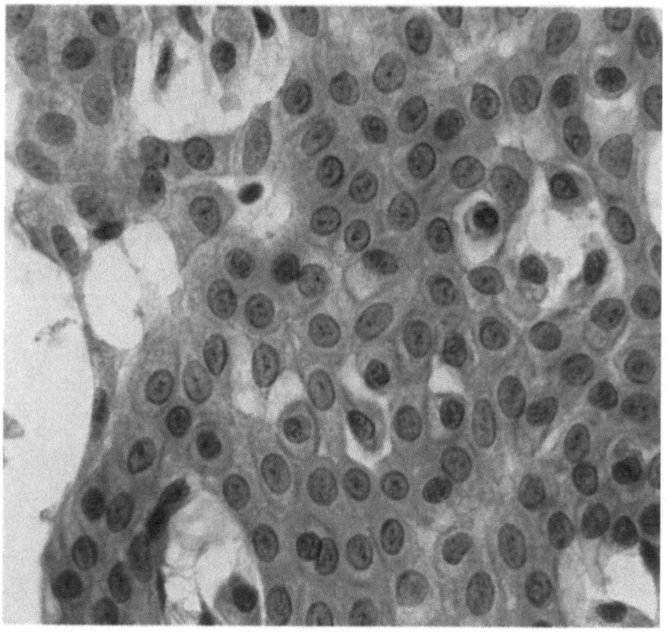

Fig. 52. Large sheet of rectal mucosal cells, in part with a honeycomb pattern like that of prostatic epithelium. The thick nuclear membranes, the loose chromatin structure and the predominantly round to oval nuclear shape nevertheless provide definite evidence that these are rectal mucosal cells. ×400

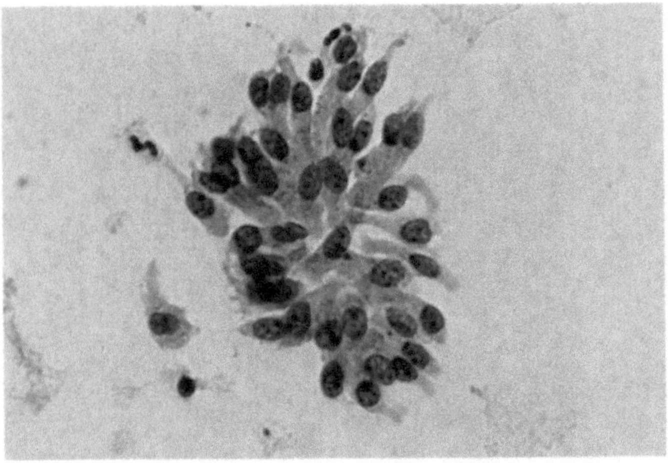

Fig. 53a. Sheet of urothelial cells from the deep layer (parabasal cells) with typical tail-like cytoplasmic extensions and round to oval, basally situated nuclei. ×400

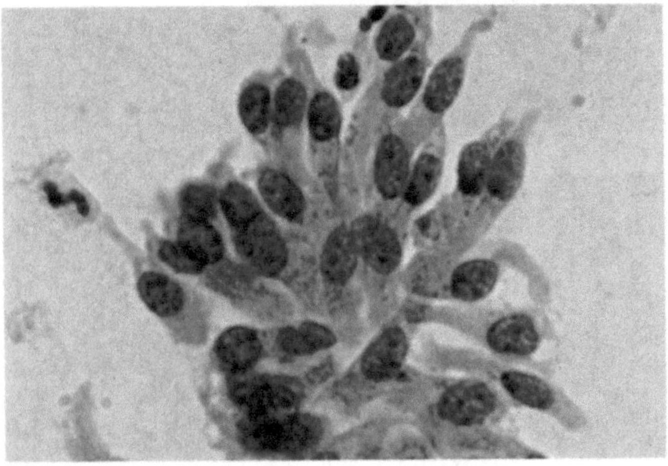

Fig. 53b. Same case at higher magnification: moderately condensed chromatin and conspicuous nucleoli; no indication of malignancy. ×630

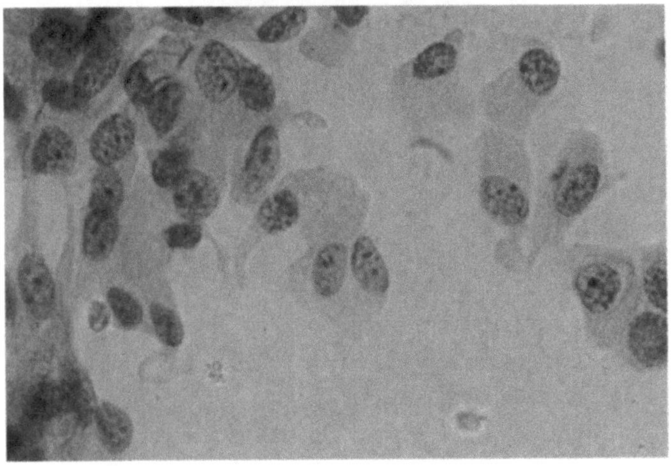

Fig. 54. Loose group of urothelial cells from the upper layer (superficial cells): ample cytoplasm, in part with two nuclei; loose chromatin; conspicuous but not prominent nucleoli. ×400

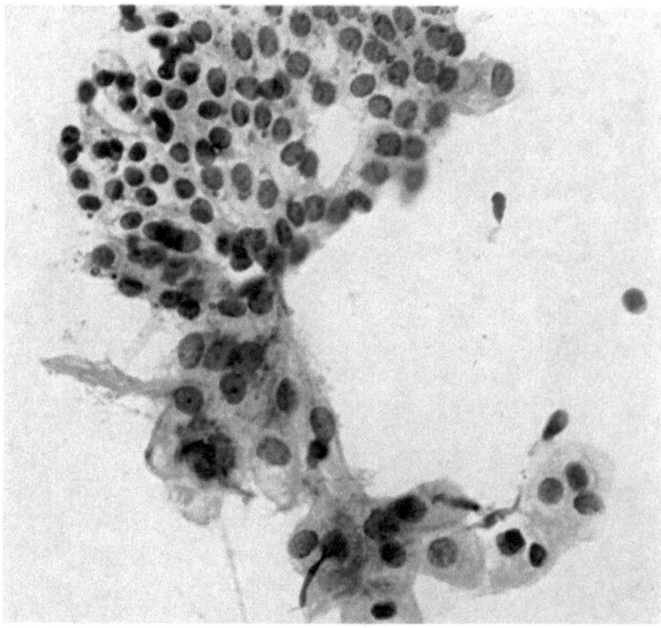

Fig. 55a. Sheet of prostatic epithelium with metaplastic squamous cells in the lower part of the picture. ×400

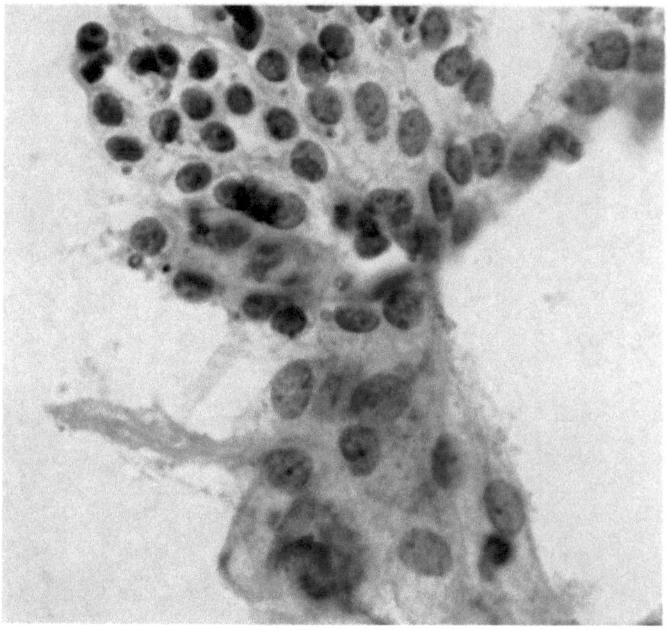

Fig. 55b. Same case at higher magnification. The nuclei of the metaplastic epithelial cells are larger than those of the prostatic cells. Slightly conspicuous nucleoli; cyanophilia of the cytoplasm. ×630

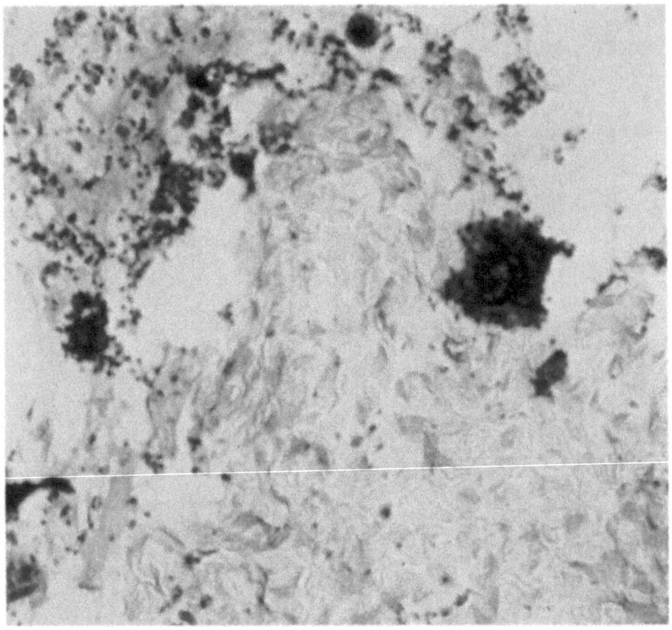

Fig. 56a. Collection of sheets of keratin

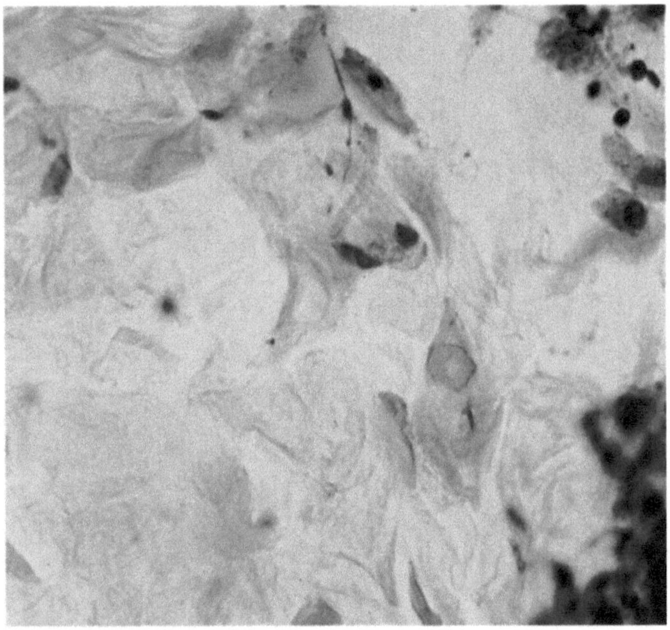

Fig. 56b. Same case at higher magnification. ×400

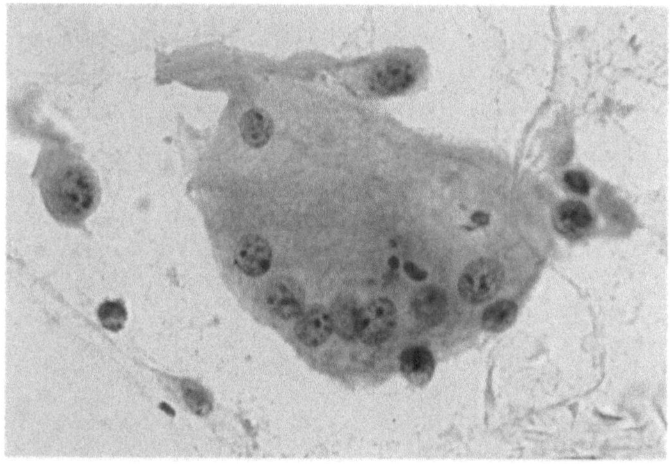

Fig. 57. Large, multinucleated histiocyte of the Langhans' type, with phagocytized nuclear debris. ×400

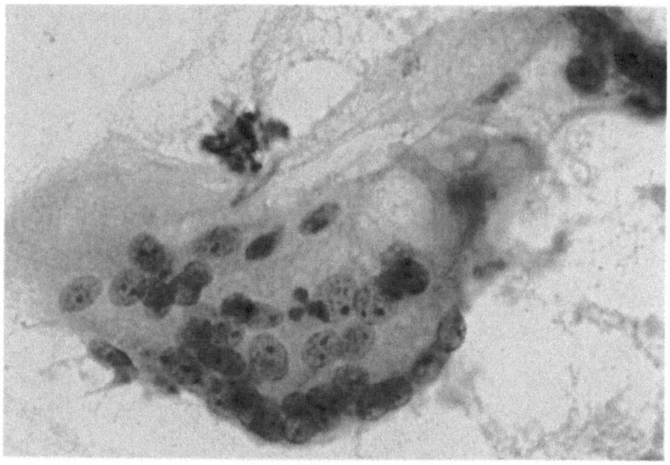

Fig. 58. Large, multinucleated histiocyte. The nuclei contain strikingly conspicuous to slightly prominent nucleoli but loosely granular chromatin and sharp definition of all nuclear membranes (cf. Fig. 57). ×400

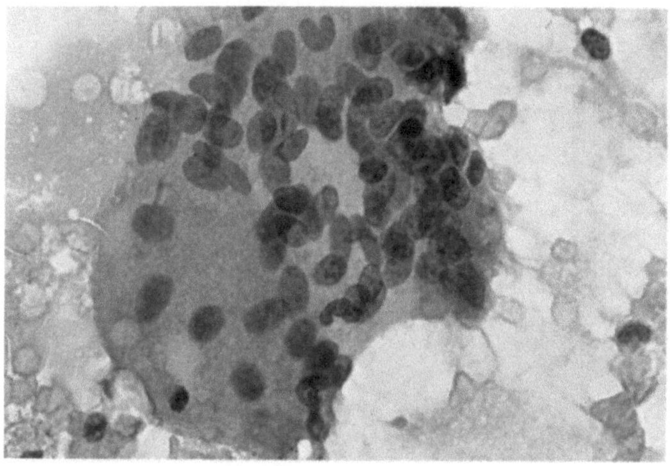

Fig. 59. Multinucleated, in part autolytically altered histiocyte, recognizable on account of the very clearly defined nuclear membranes and the richly developed cytoplasm. ×400

6 Artefacts

The most frequent artefact is *autolysis of the cellular material.* When spray or wet fixation methods are employed, prior to staining with Papanicolaou's stain, fixation must be completed within a few seconds (!) of smearing if even the slightest autolytic changes are to be avoided (see p. 12).

Consequences of poor fixation:

- The *cytoplasm* becomes considerably more pale and thin and filmy. The cell borders are blurred or no longer detectable **(Fig. 60).** If the cytoplasm has completely disappeared, only the naked nuclei are still demonstrated **(Fig. 61).**
- The *nuclei* swell as a result of autolysis and become correspondingly larger; they become irregular in shape, some of them polygonal **(Figs. 60, 61).** The chromatin is apparently homogeneous and without structure; it may even aggregate **(Fig. 60b).**
- The *nucleoli* can no longer be evaluated **(Fig. 60b).** The substantial changes in the nucleus which accompany pronounced autolysis make it impossible to arrive at even a tentative diagnosis. Only the *arrangement of the nuclei* and their *density* may provide evidence that the group of cells does not, at least, consist of normal epithelium.

In the evaluation of smears considerably altered by autolysis, it is vital to recognize that reliable assessment of the material will no longer be possible and that a further aspiration biopsy is therefore to be recommended!

If autolytic changes are only moderate, however, it is still possible to differentiate cells displaying the classical structure, e.g. multinucleated histiocytic giant cells (macrophages, **Fig. 59**), or epithelial cells from the rectal mucosa; it is on account of their shape and the basally situated nuclei **(Fig. 62)** that the latter can sometimes be assigned with confidence to the rectal mucosa, though in other cases this is problematic **(Fig. 63).**

However, nuclei from treated carcinomas *cannot be evaluated or distinguished* from histiocytes *with any reliability* even when autolysis is only moderate **(Figs. 64, 65).**

Urothelial cells from the deep layer can be identified in the presence of slight autolysis even when they are not optimally preserved, on the basis of the characteristic contour of their cytoplasm and the basal or central location of their nuclei **(Fig. 67).**

If excessive pressure is placed on the two slides *during smearing,* the nuclei of prostatic epithelial cells may not only be deformed but actually stretched so that linear or thread-like structures appear within the cell group **(Figs. 68, 69).** The more atypical the prostatic epithelial cells, the more pronounced are these artefacts; they are thus found predominantly in prostatic carcinoma. The sensitivity of nuclei to potential damage, such as that which may arise through smearing, increases in accordance with the grade of malignancy **(Fig. 69).** Nevertheless, such artefacts have scarcely any diagnostic influence since they are almost exclusively local.

Differential diagnosis must take spermatozoa into consideration, though these can be distinguished beyond all doubt by virtue of their easily recognizable heads and the sheets of seminal vesicle cells that can also be demonstrated in most instances.

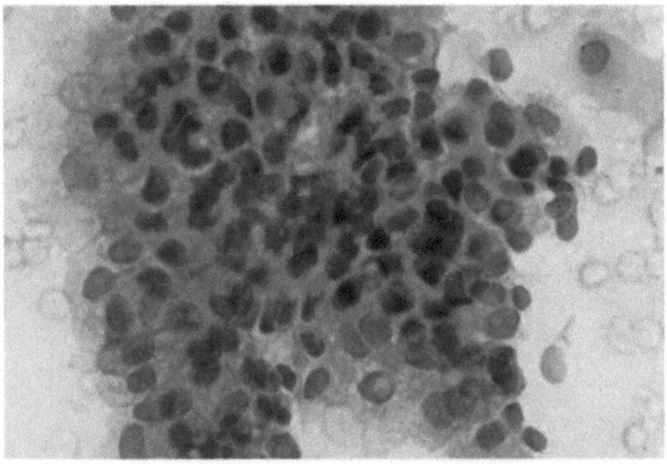

Fig. 60 a. Autolysis of prostatic epithelium resulting from inadequate fixation. The nuclear structures can no longer be evaluated reliably. ×400

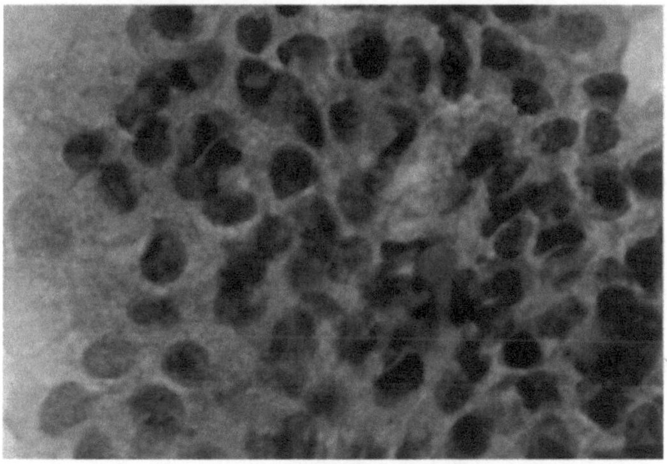

Fig. 60 b. Same case at higher magnification. ×630

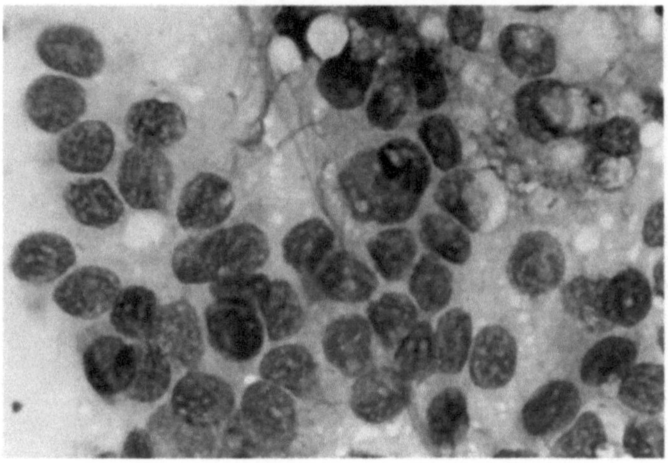

Fig. 61. MGG-stained prostatic epithelium following too brief air drying: some of the nuclei have no cytoplasm and the chromatin structure shows signs of shrinkage. ×630

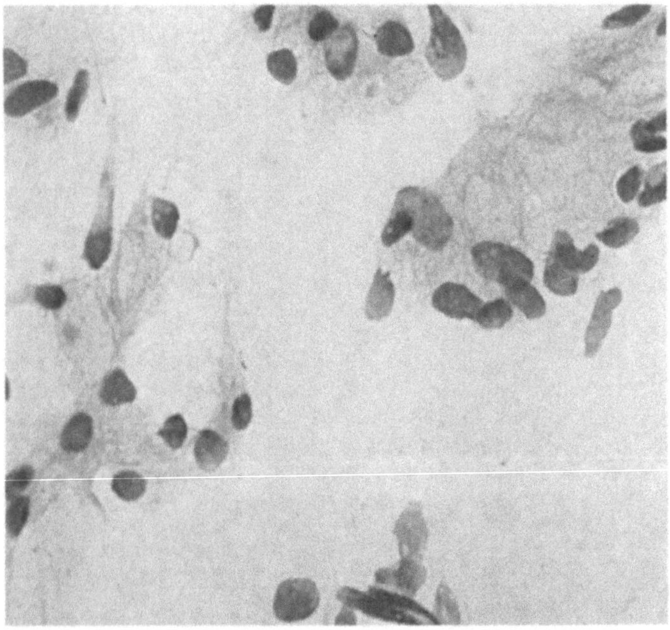

Fig. 62. Smaller groups of partially autolytic rectal mucosal cells with round to oval, basally situated nuclei and 'ballooning' of the cytoplasm. × 400

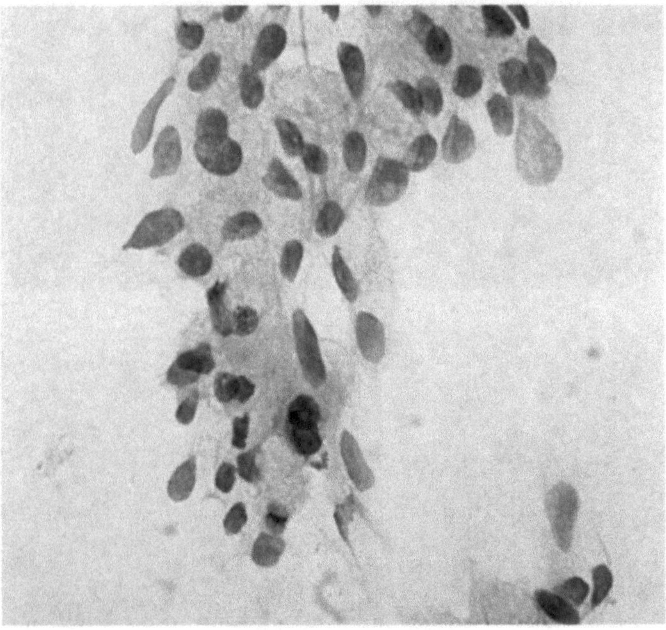

Fig. 63. Sheet of partly autolytic rectal mucosal cells, which can be classified only with great difficulty. × 400

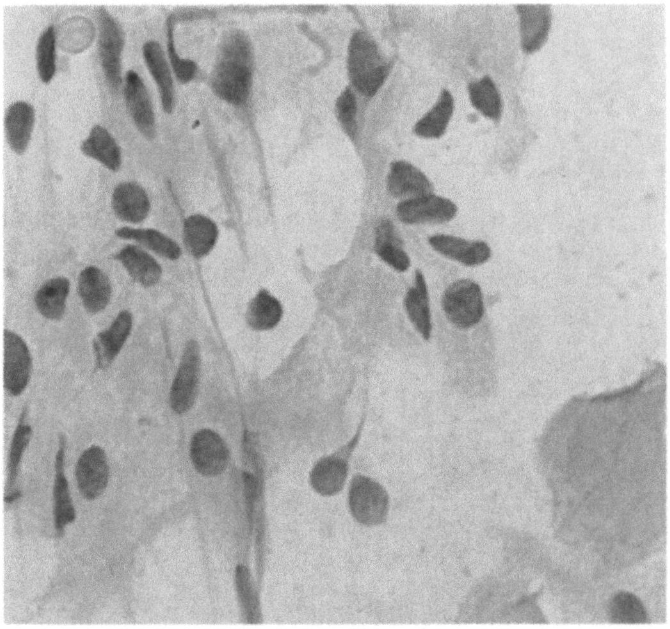

Fig. 64. Partially autolytic prostatic epithelium following hormone therapy. Owing to the fact that the nuclear structures cannot be assessed, it is not possible to establish the regression grade. ×400

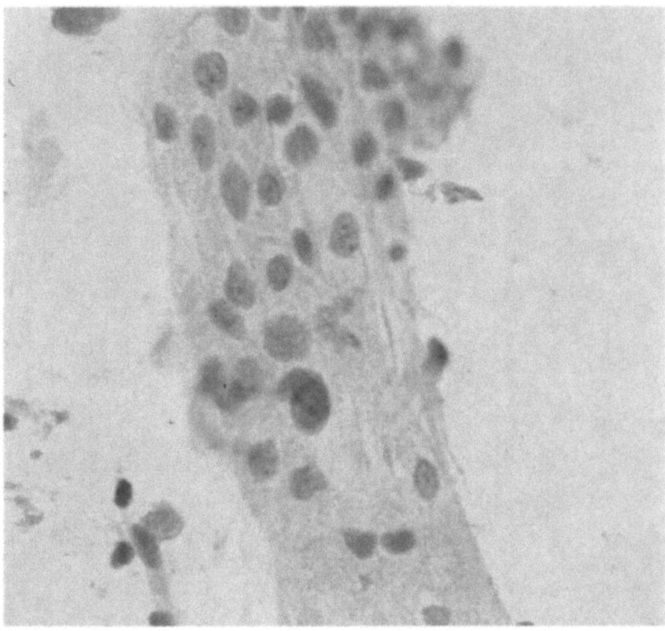

Fig. 65. Partially autolytic group of cells from a prostatic carcinoma treated with hormones: definite differential diagnosis is not possible between a group of carcinoma cells having undergone regressive changes with marked rarefaction of the nuclei, and a multinucleated histiocytic giant cell. ×400

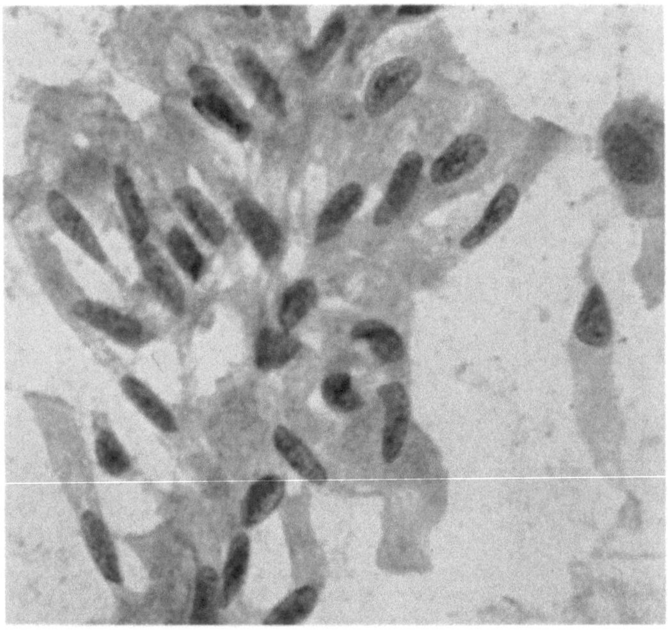

Fig. 66. Superficial autolytic changes in a sheet of urothelial cells from the deep layer: round to oval nuclear shape, basal location of the nuclei, tail-like cytoplasmic extensions. ×400

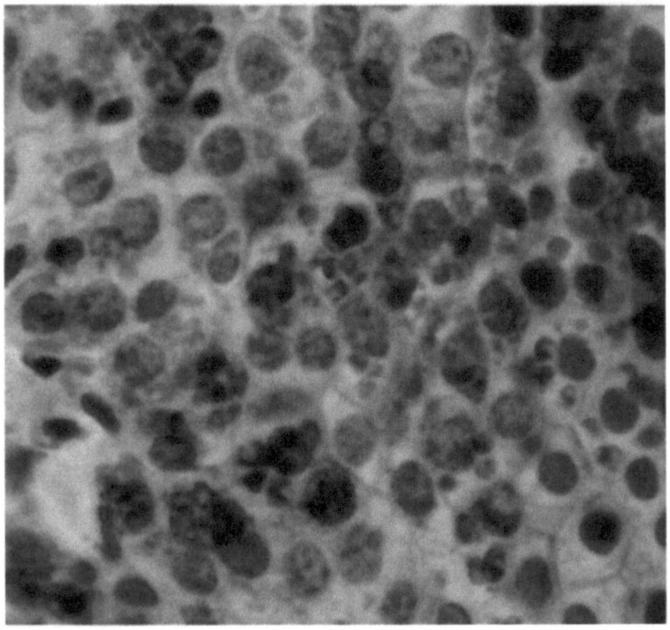

Fig. 67. Abundant intracytoplasmic granules in prostatic epithelial cells. ×630

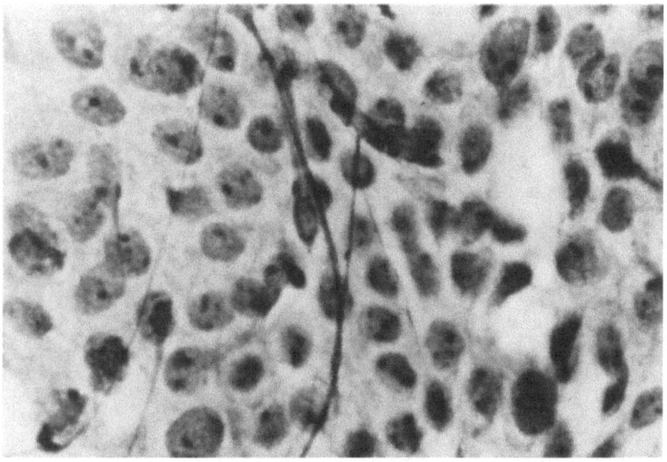

Fig. 68. Crushing artefacts: long, thread-like nuclear projections in grade III prostatic carcinoma, which, however, do not impair diagnosis. ×400

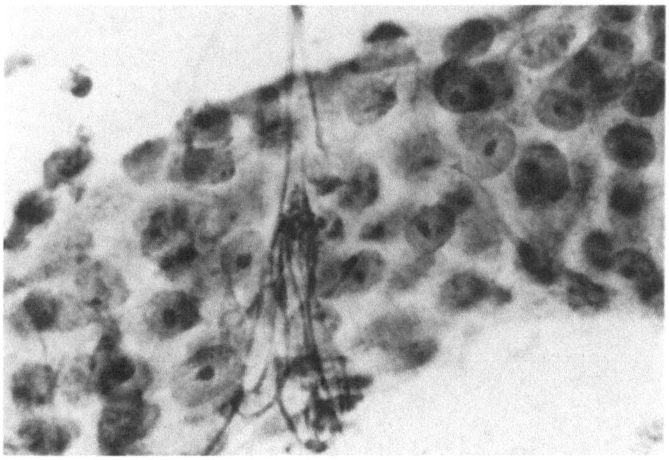

Fig. 69. Crushing artefacts: grade III carcinoma with numerous short, thread-like nuclear projections. Diagnosis is not impaired. Oil immersion, ×540

7 Primary Diagnosis of Carcinoma

In the primary diagnosis of prostatic carcinoma, cytology has two functions to fulfil, namely

- *to secure the diagnosis,* and
- *to establish the grade of malignancy.*

7.1 Procedural Reliability

Whether a prostatic carcinoma can indeed be confirmed cytologically when palpatory findings are suspicious depends upon the clinician's experience in assessing the prostate by palpation and upon the operator's mastery of the aspiration technique.

This has been illustrated recently by the results achieved by ESPOSTI (1982) in collaboration with cytologists and experienced urologists:

- When palpatory findings were *'typical'* of carcinoma, the clinical diagnosis was confirmed cytologically in 98% of cases.
- When palpatory findings were *'suspicious'*, carcinoma was proven in 68% of cases.
- When palpatory findings were simply *'atypical',* carcinoma was established in only 17% of cases.

This correspondence between positive results of palpation and cytological tests, achieved under optimal conditions, contrasts with the results of large series in which carcinomas were confirmed cytologically in an average of only 27% (17.4%–44.9%) of aspirates when they had been suspected at the clinical examination or palpation had revealed apparently typical signs. Of the other findings, 54% (29–62%) were *normal* and 6% (3.7 – 10%) *gave reason to suspect carcinoma* **(Table 7).**

The findings of the remaining biopsies in these series, which are not separately presented, relate above all to a not inconsiderable number (20%) of inflammatory changes (p. 116) and to smears that could not be evaluated owing to insufficient or poorly prepared material (0.7–60%) **(Table 1,** p. 3).

Using *punch biopsy,* by contrast, in the series shown in **Table 8** carcinoma was established histologically in an average of 39.6% (26–52%) of cases in which the palpatory findings had been suspect; however, the number of histologically 'suspect' or 'inflammatory' findings was not given in most of these series. From this it would seem that punch biopsy is superior to aspiration biopsy; this apparent superiority is in fact based upon several factors:

- Because of the low complication rate and the minimal discomfort to the patient, aspiration biopsy is doubtlessly often performed even when the findings at palpation are only very slightly suspect. This explains the unusually high proportion of the diverse forms of prostatitis.
- Not only is aspiration biopsy technically more difficult, but cytological evaluation can also be frustrated by incorrect processing (smearing and fixation) of the specimen **(Table 1).** This source of error is absent when punch biopsy is technically success-

Table 7. Results of primary cytological diagnosis of cases in which carcinoma was suspected at palpation (representative series)

Author(s)	Year	*n*	Cytology					
			Normal findings		Suggestive of carcinoma		Carcinoma	
			n	(%)	*n*	(%)	*n*	(%)
FAUL	1975	1382	401	(29.0)	138	(10.0)	358	(26.0)
DROESE et al.	1976	288	200	(69.4)	19	(6.6)	50	(17.4)
EPSTEIN	1976	118	62	(52.5)	3	(2.6)	53	(44.9)
ACKERMANN and MÜLLER	1977	645	386	(59.8)	36	(5.6)	179	(27.7)
BISHOP and OLIVER	1977	182	113	(62.1)	–	–	37	(19.7)
MELOGRANA et al.	1982	87	49	(56.3)	–	–	19	(21.8)
ESPOSTI	1982	4630	2780	(60.0)	325	(7.0)	1410	(30.5)
Personal material	1982	1086	499	(46.0)	40	(3.7)	315	(29.0)

ful, since no further processing of the specimen is necessary.

- Conversely, punch biopsy is considerably more discomforting and has a much higher complication rate; therefore in the urological practice, at least, it is performed only when suspicion of prostatic carcinoma is well grounded.
- If the indications for punch biopsy are extended to include less suspect findings, the *rate of findings positive for carcinoma* is only 22% (33/149); this corresponds – in the presence of a high complication rate – to the rate achieved with aspiration biopsy or *even lies below it* (BISSADA et al. 1977).

As assessed by the histological results of punch biopsy, the *reliability of aspiration biopsy* in cytological primary diagnosis of prostatic carcinoma averages 82% (47.4–96%) **(Table 9)**, while that of *punch biopsy* is reported as being 80–90% (BISSADA et al. 1977).

Since a cytologically or histologically negative biopsy does *not* exclude a carcinoma, biopsy must be repeated if clinical findings continue to evoke suspicion of carcinoma. Renewed aspiration biopsy provides cytological confirmation of clinically suspected carcinoma in 7.8% (2–14%) of cases on average **(Table 10)**. The corresponding results with a second punch biopsy are reported to average 6.3% (GRAYHACK and BOCKRATH 1981).

Table 8. Frequency of histologically proven carcinoma, following suspect findings at palpation, in large series of transrectal and transperineal punch biopsies

Author(s)	Year	*n*	Positive histology	
			n	(%)
KAUFMAN and SCHULTZ[a]	1962	656	250	38
FORTUNOFF[a]	1962	286	75	26
SIKA and LINDQUIST[a]	1963	300	129	43
BARNES and NINAN	1972	217	78	36
ZINCKE et al.	1973	342	177	52
ACKERMANN and MÜLLER[a]	1977	642	235	37
BISSADA	1977	306	151	49
LEISTENSCHNEIDER and NAGEL	1978	977	351	36

[a] Transperineal punch biopsies

Table 9. Sensitivity of prostatic cytology in the primary diagnosis of carcinoma in cases of histologically confirmed carcinoma

Author(s)	Year	Sensitivity (%)
ESPOSTI	1966	89.6
EKMAN et al.	1967	90.2
KAULEN et al.	1973	96.2
FAUL	1975	93.0
DROESE et al.	1976	47.4
EGLE et al.	1976	97.5
BISHOP and OLIVER	1977	74.0
ACKERMANN and MÜLLER	1977	68.9
LEISTENSCHNEIDER	1981	85.6
MELOGRANA et al.	1982	80.0
ESPOSTI	1982	94.0

Table 10. Increase in the sensitivity of prostatic cytology achieved through repeat biopsy

Author(s)	Year	Positive at second biopsy
FAUL	1975	14 %
DROESE et al.	1976	6.4%
ACKERMANN and MÜLLER	1977	2.0%
ESPOSTI	1982	9.1%

As shown by the literature and our own personal observations, it is not seldom the case that cytological confirmation of a carcinoma is already possible at *primary aspiration,* whereas in the same patient *a second or third punch biopsy* is required before carcinoma is established (DROESE et al. 1967; EKMAN et al. 1967).

In 140 cytologically and histologically proven carcinomas, FAUL (1975) was able to secure the diagnosis at primary investigation in 83% of cases *cytologically* but in only 70% *histologically* by means of punch biopsy. We, ourselves, demonstrated carcinoma *cytologically* in the material obtained at the initial aspiration in 11 of 12 cases, while in these 11 patients a second or third punch biopsy was necessary before *histological confirmation* was obtained.

For clinical practice, the rate of false-positive cytological diagnoses is of far greater significance than failure to demonstrate carcinoma at the first attempt.

The *rate of false-positives* is, at 0.8% (1/113; BISHOP and OLIVER 1977), 2% (5/210; ESPOSTI 1982) and 0.18% (2 of 1086 personal cases), very low; nevertheless, the results require closer scrutiny.

The apparently false-positive cytological results (5/210) published by ESPOSTI (1982) were based upon comparison with histological investigation of prostatic resection specimens (four cases) or perineal punch biopsy (one case). *In none of the five cases, however, did the subsequent clinical course contradict the primary, cytologically positive diagnosis of carcinoma!*

Reliable pronouncements on the risk of false-positive cytological results are only possible when comparison is made with specimens obtained by way of radical prostatectomy. Unfortunately scarcely any such comparative investigations have been undertaken to date.

ESPOSTI (1982) reported that histological examination after radical prostatectomy confirmed the *exclusively* cytologically based diagnosis of locally delimited carcinoma in each of 18 patients. Similarly, in two patients with positive preoperative aspiration biopsy results – but negative punch biopsy results (!) – KAUFMAN et al. (1982) confirmed the cytological diagnosis histologically after radical prostatectomy.

To summarize, the results published to date permit us to define the reliability of aspiration biopsy for the primary diagnosis of prostatic carcinoma as follows:

- A clear-cut positive cytological diagnosis in the presence of a simultaneously negative histological result can no longer be 'dismissed' as false-positive when the aspirate has been correctly obtained and processed, and the person performing the evaluation is experienced in cytodiagnosis of the prostate.

- A cytologically suspect or negative finding requires renewed biopsy if the clinical appraisal continues to give reason to suspect carcinoma. This also applies to negative primary punch biopsy results under the same circumstances!

7.2 Cytological Criteria for Prostatic Carcinoma

Cytological diagnosis of prostatic carcinoma depends upon the assessment of specific alterations in the *cell* itself and the *nucleus,* whereby the latter are of decisive importance.

The most important parameters for diagnosis and grading are:

structure of cell groups
arrangement of the nuclei
size and shape of the nuclei
size and shape of the nucleoli
chromatin structure
structure of the nuclear membranes

7.2.1 Structure of Cell Groups

Alterations in the structure of groups of cells can already be identified when scanning the smear at low magnification (100-fold). Generally, the margins of the groups are no longer smoothly bounded; rather, in contrast to normal findings or slight to moderate atypia, varying degrees of dissociation of cells are seen. Correspondingly, the cell borders can be recognized only in part or even not at all, and the cytoplasm is more basophilic **(Fig. 71).**

7.2.2 Alterations in the Nuclei

7.2.2.1 Arrangement

The arrangement of the nuclei can also be evaluated while examining the aspirate at the screening magnification. It is characteristic of carcinoma that the nuclear pattern is, to varying degrees, disturbed; in some of the groups of carcinoma cells at least, the nuclei no longer lay side by side in an orderly fashion **(Figs. 71, 75, 81, 82).**

In contrast to the *monolayering* and *regular spacing* of the nuclei in a normal sheet of cells, the disturbance of the structure of malignant groups of cells frequently leads to *overlapping of the nuclei.* The higher the grade of malignancy, the less well arranged are the group of cells and the nuclei **(Figs. 84, 85, 104).**

A pseudo-acinar arrangement of the nuclei can result in the picture of so-called microadenomas, which are typical, though not specific, for well differentiated carcinoma (Esposti 1966, 1982; Faul 1975; Staehler et al. 1975) and which in our experience are observed more frequently in MGG-stained smears after air drying than in spray-fixed, Papanicolaou-stained smears **(Figs. 73, 74).**

When these obvious structural alterations are observed during preliminary screening, close examination of the entire smear at a magnification of 400- to 500-fold is absolutely essential, since in a Papanicolaou-stained smear it is only at this magnification that those nuclear structures decisive for the diagnosis of carcinoma can be reliably evaluated.

Because of the high transparency of Papanicolaou's stain, optimal differentiation of precisely the cytologically significant nuclear parameters is possible.

7.2.2.2 Size and Shape

The nuclei are invariably larger than in normal **(Fig. 29a)** or mildly to moderately atypical cells **(Figs. 70, 71a).** Since normal sheets of epithelial cells are always also visible in cases of less extensive carcinoma, diagnosis is facilitated by comparing the various nuclear sizes.

In most cases the *nuclear size* varies to a greater or lesser extent **(Figs. 71a, 72, 82).** As the grade of malignancy increases, so one encounters true *nuclear polymorphism* with lobulation of the nuclei. In addition, the nuclei are more fragile, so that crushing artefacts are frequent **(Figs. 105, 109).**

The normal *round nuclear shape* also alters with increasing dedifferentiation, through *oval* and *triangular* to *polygonal* forms **(Figs. 105b, 108, 110).**

Mitoses are rare, but are more often observed in treated prostatic carcinoma with increasing resistance to therapy. They are *not* a criterion for the primary diagnosis **(Fig. 107).**

7.2.2.3 Nucleolus

The nucleoli are always *prominent!* With increasing malignancy, however, their shape and number change. While in well-differentiated carcinomas nucleoli are essentially *round* and each nucleus contains just one or two **(Figs. 70, 71),** *loss of circularity* **(Figs. 86b, 87)** (wedge-shaped, triangular or pencilliform) and an *increase per nucleus* **(Figs. 105b, 107)** are indicative of a higher or a high grade of malignancy.

7.2.2.4 Chromatin

The nuclear chromatin is always condensed and with increasing dedifferentiation displays an irregular distribution, often with clumping. In consequence of this increased density, the nuclear transparency characteristic of normal findings after Papanicolaou staining is *no longer* to be seen **(Figs. 71b, 89b).**

7.2.2.5 Nuclear Membrane

At 500-fold magnification the nuclear membrane can often be demonstrated only incompletely, while it is generally not to be seen at all in high-grade malignancies. Only in grade I carcinoma may it be intact in most nuclei **(Figs. 81, 82).**

It is decisive for the diagnosis of prostatic carcinoma that all the cited parameters are demonstrated within the pathological groups of cells. Foci of prominent nucleoli, variations in nuclear size and increases in chromatin density within otherwise normal sheets of prostatic epithelium alone are no proof of carcinoma.

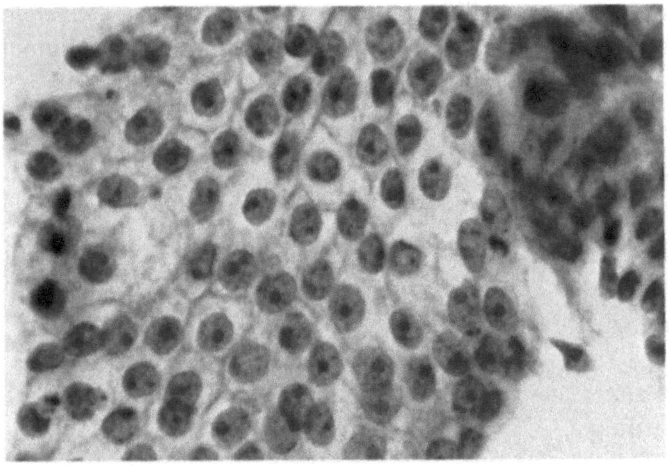

Fig. 70. Prostatic carcinoma with the typical characteristics: condensed nuclear chromatin, prominent nucleoli, enlarged nuclei. Nuclear polymorphism and disturbance of nuclear arrangement are still slight. ×400

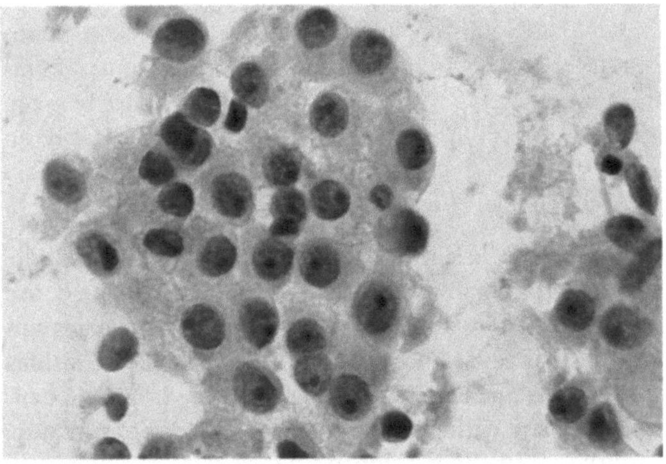

Fig. 71a. Two groups of prostatic carcinoma cells showing slight dissociation locally and somewhat more pronounced nuclear polymorphism and disturbed nuclear arrangement. Some nuclear membranes are still recognizable. ×400

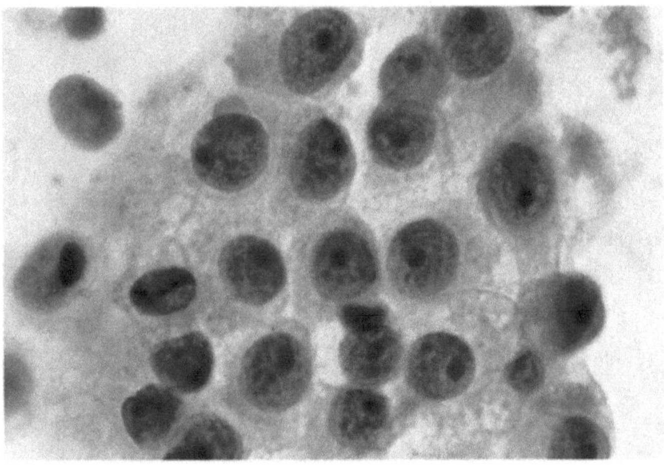

Fig. 71b. Same case at higher magnification: marked chromatin condensation with a somewhat clumped chromatin structure in places; some of the very prominent nucleoli are beginning to lose their circular contour. Oil immersion, ×1000

Fig. 72. Group of prostatic carcinoma cells alongside faded erythrocytes: marked nuclear polymorphism, condensation of the chromatin, prominent nucleoli. In places there is more than one nucleolus per nucleus. Marked disturbance of the nuclear arrangement. ×630

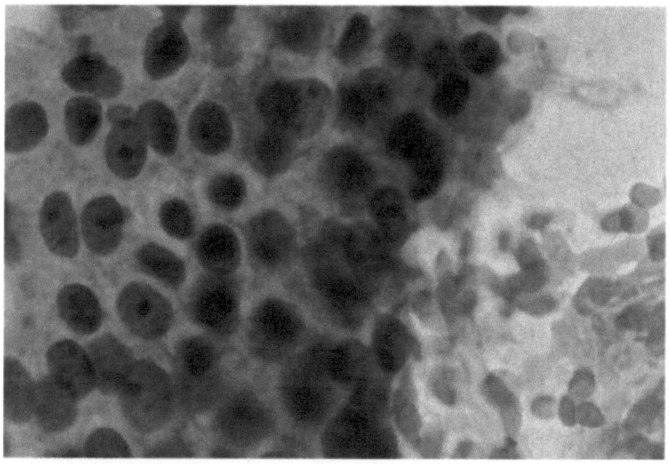

Fig. 73. Group of prostatic carcinoma cells, stained according to May-Grünwald-Giemsa: moderate nuclear polymorphism and disturbance of nuclear arrangement, with prominent nucleoli throughout. A so-called micro-adenoma is visible in the center. ×400

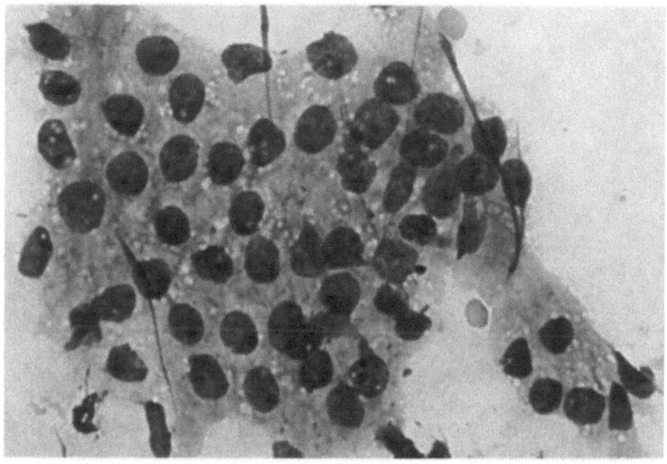

Fig. 74. Same group of cells at higher magnification. ×630

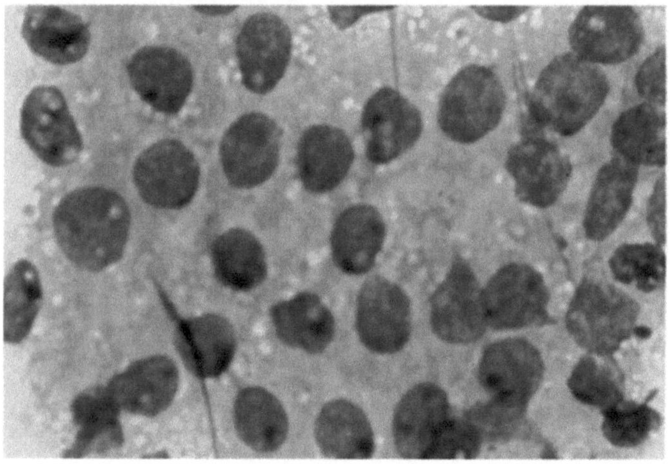

8 Grading of Prostatic Carcinoma

8.1 Histology

Classification of prostatic carcinoma according to the criteria proposed by DHOM (1980) takes into consideration structural characteristics and depends to a large extent upon the architecture of tumoral growth (glandular and stromal features). Distinction is to be drawn between prostatic carcinomas having a *uniform* structural pattern and those whose structural pattern is *pluriform*, the latter predominating at 54%. Special forms of prostatic carcinoma are very rare, accounting for only 2.3% of cases **(Table 11).**

By combining these morphological criteria with the criteria pertaining to anaplasia it is possible to arrive at reproducible and prognostically reliable histological grading.

Grading may be defined as the classification of tumors into groups with different malignancy potentials on the basis of histological or cytological criteria.

Although it is possible to grade carcinomas solely by reference to their *glandular patterns* (GLEASON 1966), the interindividual reproducibility is, at 67%, unsatisfactory (HARADA et al. 1977).

In *histological* grading also, evaluation of the nucleus is of crucial significance for the prognosis and thus for the validity of the grading (MOSTOFI 1966; HARADA et al. 1977; DHOM 1980; BÖCKING 1980, 1982; GAETA 1981).

Most authors classify prostatic carcinomas into three different grades of malignancy, designated G I to G III.

The parameters suggested by BÖCKING (1981) for histological grading with three grades of malignancy were in essence adopted by the *"Uropathological Study Group on 'Prostatic Carcinoma'"*[1] (MÜLLER

Table 11. Histological classification of prostatic carcinoma (DHOM 1980)

	n	(%)
I. Common prostatic carcinoma		
A. Uniform structural pattern		
1. Well differentiated adenocarcinoma	924	13.67
2. Poorly differentiated adenocarcinoma	1057	15.64
3. Cribriform Ca	470	6.95
4. Undifferentiated, solid Ca	526	7.78
B. Pluriform structural pattern		
1. Well and poorly differentiated adenocarcinoma	463	6.85
2. Cribriform and solid Ca	408	6.04
3. Cribriform pattern in other types	1707	25.26
4. Other combinations	1046	15.48
II. Special forms		
1. Endometrioid Ca	8	0.12
2. Urothelial (transitional cell) Ca	126	1.86
3. Squamous cell Ca	14	0.21
4. Mucus-forming Ca	9	0.13
	6758	100.00

Acc. to DHOM, Deutsches Prostatakarzinomregister (German Register of Prostatic Carcinomas), Homburg (Saar), W. Germany

1 Pathologisch-Urologischer Arbeitskreis 'Prostatakarzinom'

Table 12. Histological grading of prostatic carcinoma (Pathologisch-Urologischer Arbeitskreis "Prostatakarzinom" 1980; Böcking 1981)

Grade of differentiation		Grade of nuclear anaplasia		Definition of grade according to score	
Well differentiated	0	Slight	0	Grade I	Score 0–1
Poorly differentiated	1	Moderate	1	Grade II	Score 2–3
Cribriform	2	Severe	2	Grade III	Score 4–5
Solid	3				

et al. 1980) **(see Table 12).** In this system the rating number denoting the glandular differentiation is added to that representing the grade of nuclear anaplasia. The sum of these numbers (the 'score') then indicates the grade of malignancy (G I to G III). The interindividual reproducibility of the system is 91% (Bocking et al. 1982).

8.2 Cytology

Owing to the nature of the material to be examined, *cytological grading* cannot be based upon the criteria of glandular structure and stroma; rather it must depend exclusively upon specific changes at the cellular and nuclear level.

As with the classification of atypias, various grading systems are in use (Faul 1975; Voeth et al. 1978; Spieler et al. 1976). That employed by Esposti (1966, 1971, 1982), which comprises three grades of differentiation, displays good prognostic validity but inadequate interindividual reproducibility (61% according to Voeth et al. 1978; 56% according to Böcking 1980).

Faul et al. (1974, 1975, 1978) distinguish between four grades of malignancy, but found significant differences in the 3-year survival rate only between grade I and grade III or IV carcinomas.

8.3 Cytological Grading According to the Uropathological Study Group on 'Prostatic Carcinoma'

This system of grading, which correlates closely with the six parameters proposed by Böcking (1980, 1981) for histological grading, has to date proven to have the greatest validity and reproducibility (Müller et al. 1980). Correlation with the prognostically relevant criteria of the pattern of DNA distribution in tumor cell nuclei in diverse types of carcinoma has shown the above-mentioned parameters to be decisive for the prognosis of prostatic carcinoma. The characteristics in question are:

average nuclear size
variability in nuclear size
average nucleolar size
nucleolar variability (size, shape, number)
disturbance of nuclear arrangement
cellular and nuclear dissociation

Nuclear polymorphism, nuclear-cytoplasmic ratio and nuclear hyperchromatism or heterochromia are all ignored, since they do not correlate with prognosis (Böcking 1981).

Table 13 shows this system of grading, which is easy to learn on account of its standardization. The interindividual reproducibility is 86% (Müller et al. 1980).

Since more than 50% of prostatic carcinomas have a pluriform structural pattern **(see**

Table 13. Cytological grading of prostatic carcinoma (Pathologisch-Urologischer Arbeitskreis "Prostatakarzinom" 1980; BÖCKING 1980, 1981)

	Slight (1)	Moderate (2)	Severe (3)
1. Average nuclear size	(1)	(2)	(3)
2. Variability in nuclear size	(1)	(2)	(3)
3. Average nucleolar size	(1)	(2)	(3)
4. Nucleolar variability (size, shape, number)	(1)	(2)	(3)
5. Disturbance of nuclear arrangement	(1)	(2)	(3)
6. Cellular and nuclear dissociation	(1)	(2)	(3)

Each of the six criteria is assessed and scored in each of the cell complexes evaluated.
The six scores are added together
A total (score)
of 6–10 corresponds to grade I
of 11–14 corresponds to grade II
of 15–18 corresponds to grade III

Table 11), i.e. various areas of the tumor display different grades of differentiation, the general rule in cytological grading is to evaluate those cell complexes with the highest grade of malignancy and to score the lesion on this basis. This practice has been adopted because it is such complexes which are decisive for establishing a prognosis.

According to the agreement reached by the Uropathological Study Group on 'Prostatic Carcinoma', a cytological smear which contains both G III and G I portions must be classified as 'prostatic carcinoma, grade III'. This diagnosis can be supplemented by the comment that portions of a grade I carcinoma could also be identified; nevertheless, the clinical relevance of such additional information is not yet known.

When electronic data processing of the findings is undertaken, the sum (or score) upon which the grade of malignancy is based should always be included in the calculations, since it has not yet been clarified whether and to what extent prostatic carcinomas of the same grade but with different scores vary as regards prognosis, e.g. grade I (score 10) versus grade I (score 6).

Including the score in the assessment of the grade of malignancy can provide the clinician with an essential therapeutic and prognostic indicator in borderline cases. For instance, a diagnosis of prostatic carcinoma grade II with a score of 14, being closer to malignancy grade III than grade II, would give a clinician a more accurate impression of a lesion and its likely course than would *the simple* diagnosis of grade II carcinoma, which with a score of 11 would probably carry a more favorable prognosis.

With this system of grading on the basis of cytological parameters and the services of two investigators with many years' experience, we ourselves have been able to achieve *intraindividual reproducibility* of 92% in 50 carcinomas with various grades of differentiation and *interindividual reproducibility* of 83% in 42 prostatic carcinomas. Most of the classifications that could not be reproduced were borderline findings between grade I and grade II or grade II and grade III.

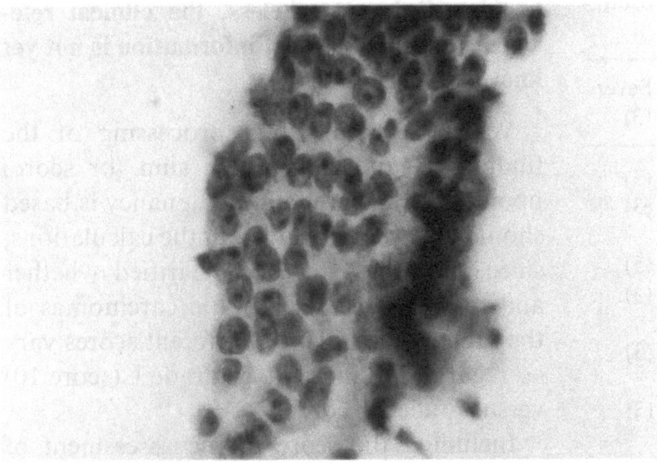

Fig. 75. Grade I carcinoma with the following evaluation of the six parameters decisive for cytological grading: average nuclear size 1; variability in nuclear size 1; average nucleolar size 1; nucleolar variability 1; disturbance of nuclear arrangement 2; nuclear dissociation 1. Score: 7. ×400

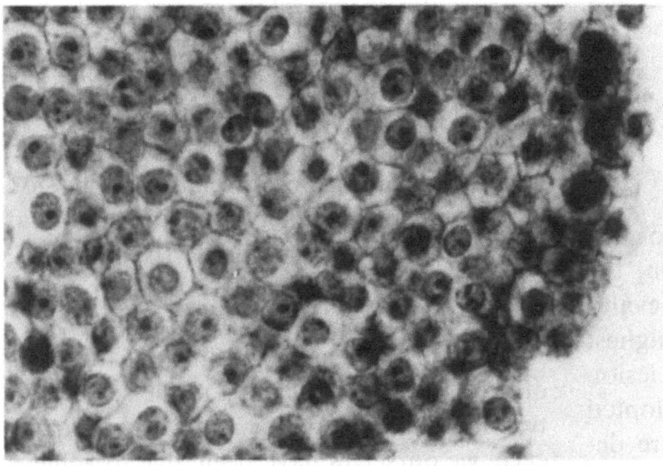

Fig. 76. Grade I carcinoma: average nuclear size 2; variability in nuclear size 2; average nucleolar size 1; nucleolar variability 2; disturbance of nuclear arrangement 2; nuclear dissociation 1. Score: 10. ×400

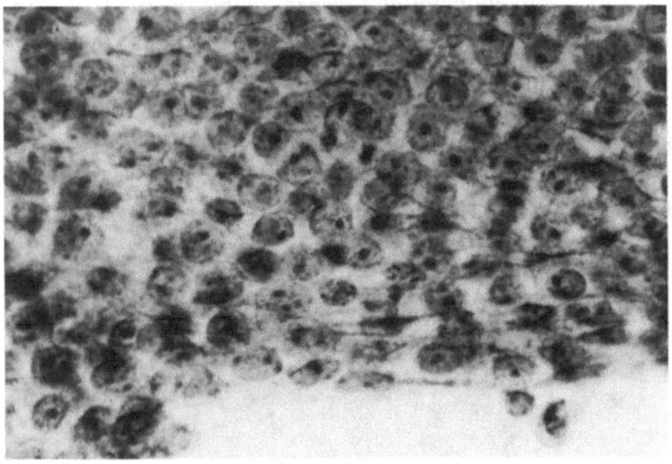

Fig. 77. Grade I carcinoma: average nuclear size 2; variability in nuclear size 2; average nucleolar size 1; nucleolar variability 2; disturbance of nuclear arrangement 2; nuclear dissociation 1. Score: 10. ×400

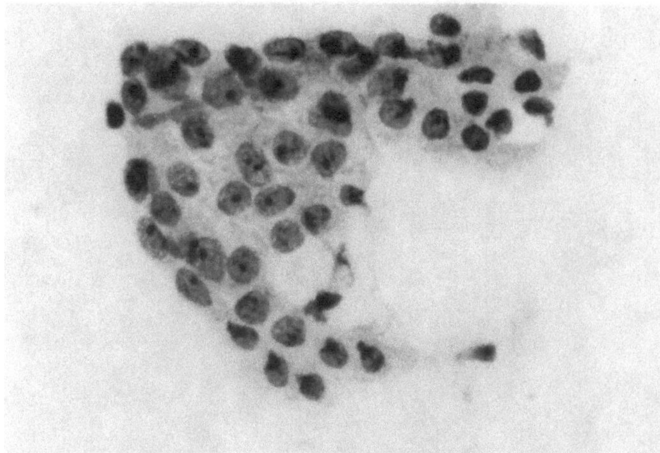

Fig. 78. Grade I carcinoma: average nuclear size 1; variability in nuclear size 1; average nucleolar size 1; nucleolar variability 2; disturbance of nuclear arrangement 2; nuclear dissociation 1. Score: 8. ×400

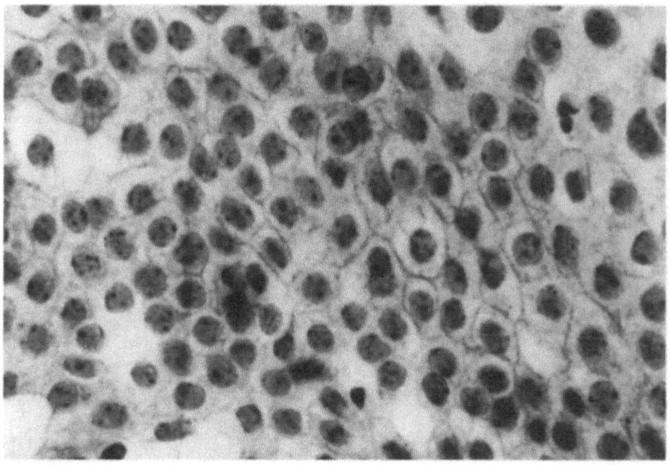

Fig. 79. Grade I carcinoma: average nuclear size 1; variability in nuclear size 1; average nucleolar size 1; nucleolar variability 3; disturbance of nuclear arrangement 2; nuclear dissociation 1. Score: 9. ×400

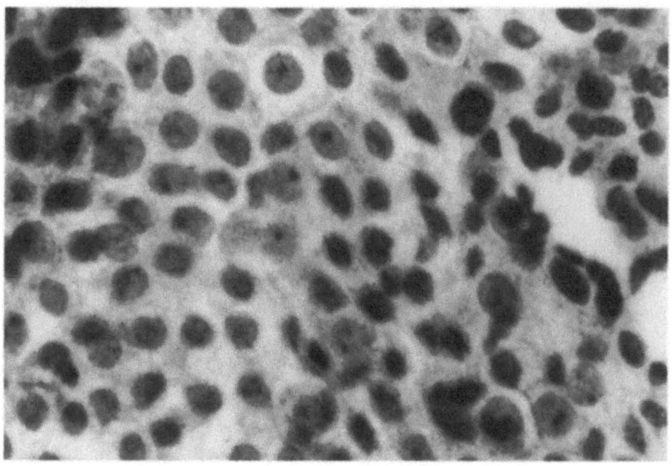

Fig. 80. Grade I carcinoma: average nuclear size 2; variability in nuclear size 2; average nucleolar size 1; nucleolar variability 1; disturbance of nuclear arrangement 2; nuclear dissociation 1. Score: 9. ×400

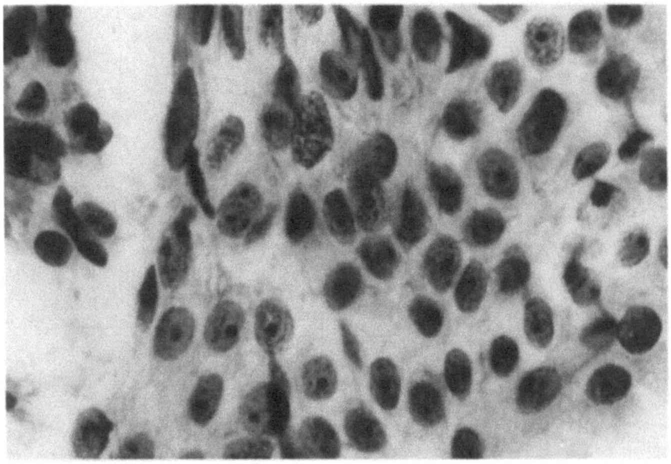

Fig. 81. Grade II carcinoma with the following evaluation of the six parameters decisive for grading: average nuclear size 2; variability in nuclear size 2; average nucleolar size 2; nucleolar variability 2; disturbance of nuclear arrangement 2; nuclear dissociation 2. Score 12. ×400

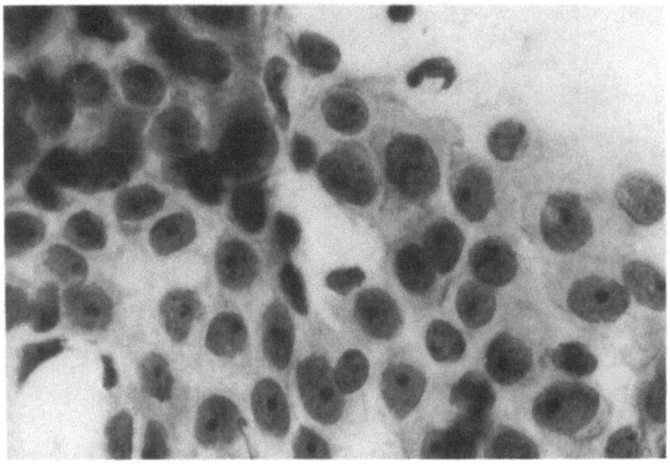

Fig. 82. Grade II carcinoma: average nuclear size 3; variability in nuclear size 3; average nucleolar size 2; nucleolar variability 2; disturbance of nuclear arrangement 2; nuclear dissociation 2. Score: 14. ×400

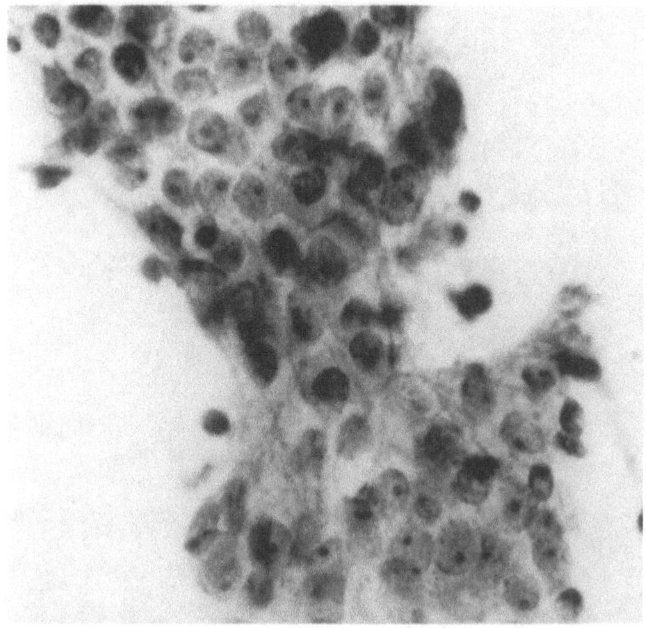

Fig. 83. Grade II carcinoma: average nuclear size 2; variability in nuclear size 2; average nucleolar size 2; nucleolar variability 3; disturbance of nuclear arrangement 2; nuclear dissociation 2. Score: 13. ×400

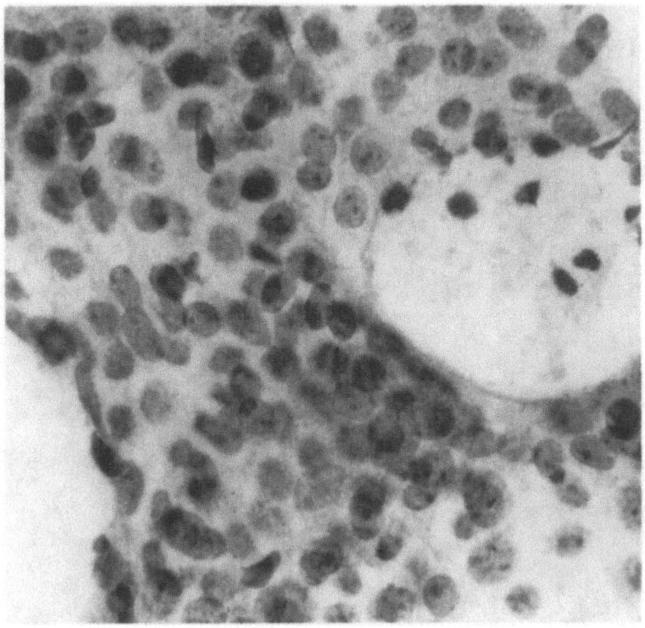

Fig. 84. Grade II carcinoma: average nuclear size 2; variability in nuclear size 2; average nucleolar size 1; nucleolar variability 3; disturbance of nuclear arrangement 2; nuclear dissociation 1. Score: 11. ×400

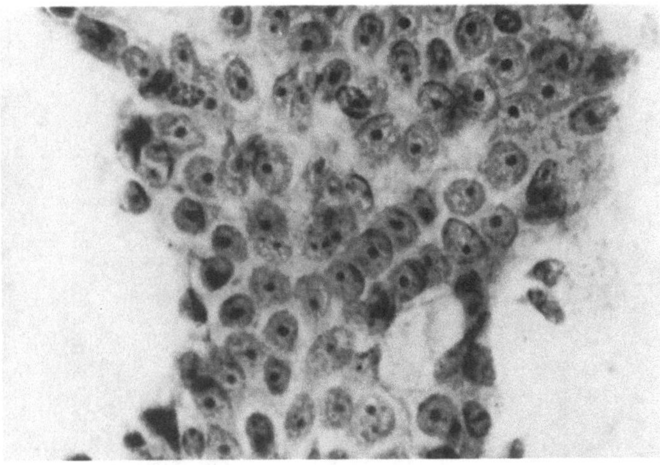

Fig. 85. Grade II carcinoma: average nuclear size 2; variability in nuclear size 2; average nucleolar size 2; nucleolar variability 2; disturbance of nuclear arrangement 2; nuclear dissociation 2. Score: 12. ×400

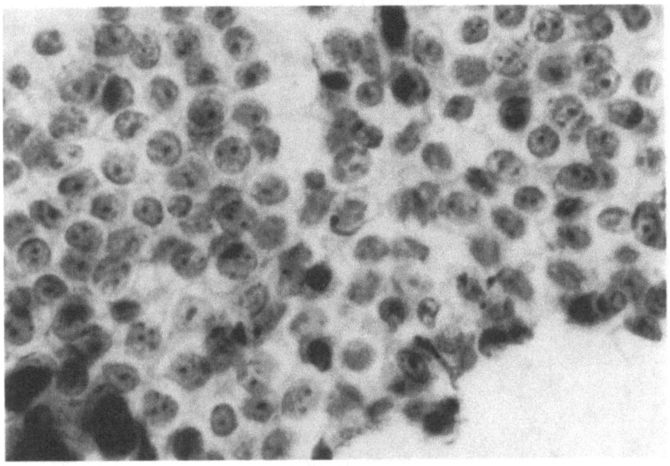

Fig. 86a. Grade II carcinoma: average nuclear size 1; variability in nuclear size 2; average nucleolar size 2; nucleolar variability 3; disturbance of nuclear arrangement 2; nuclear dissociation 1. Score: 11. ×400

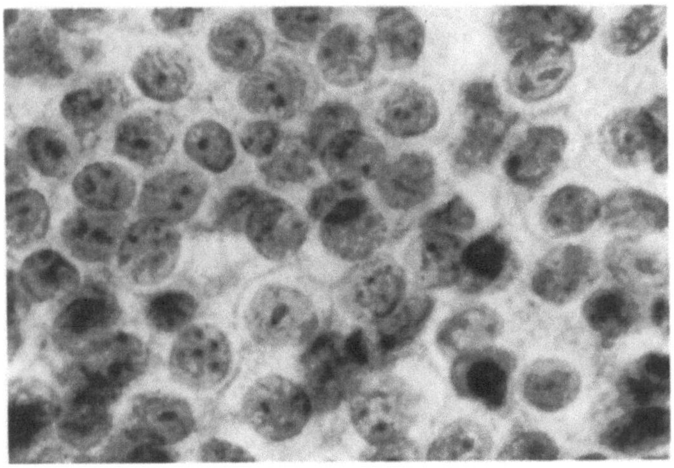

Fig. 86b. Same case at higher magnification. The considerable nucleolar variability is striking. ×630

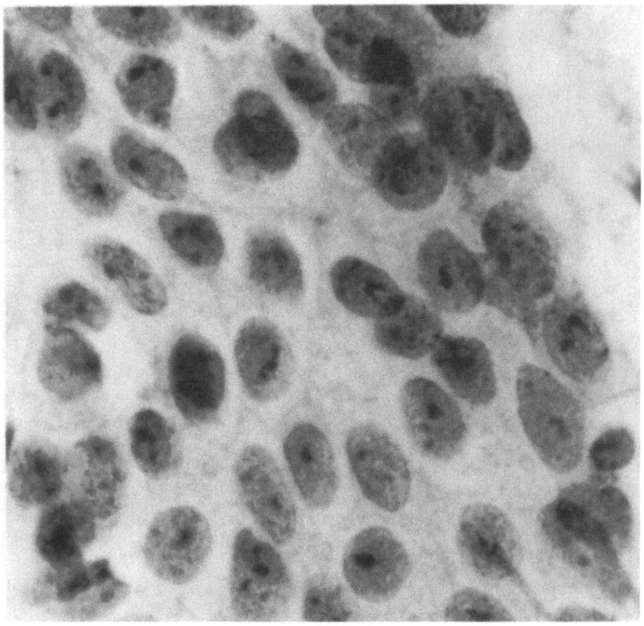

Fig. 87. Grade II carcinoma at higher magnification with the same parameters as described in Fig. 86. Score: 11. ×630

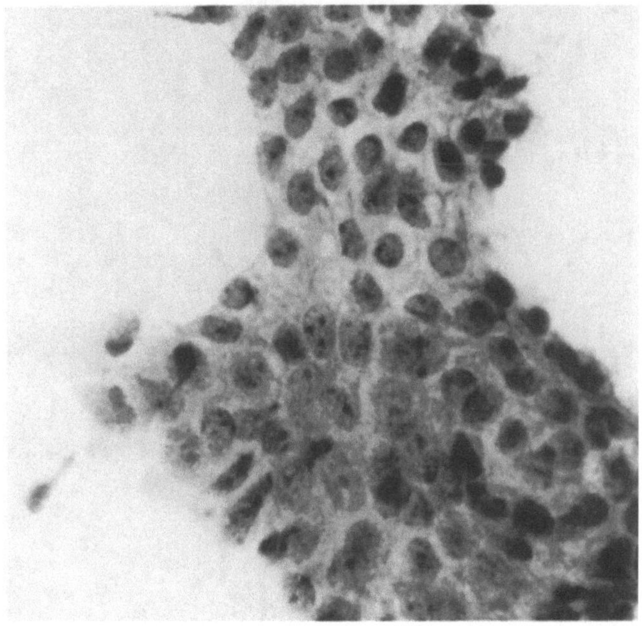

Fig. 88. Grade II carcinoma: average nuclear size 2; variability in nuclear size 3; average nucleolar size 2; nucleolar variability 3; disturbance of nuclear arrangement 2; nuclear dissociation 2. Score: 14. ×400

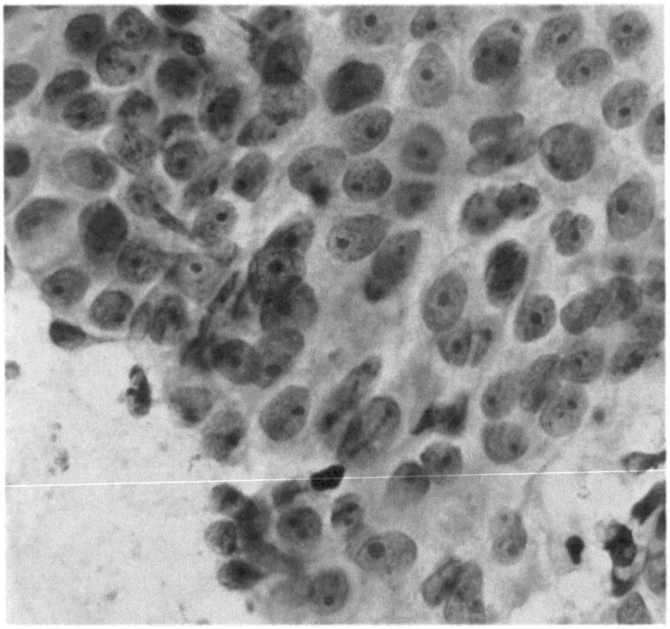

Fig. 89a. Grade II carcinoma: average nuclear size 3; variability in nuclear size 2; average nucleolar size 2; nucleolar variability 2; disturbance of nuclear arrangement 2; nuclear dissociation 1. Score: 12. ×400

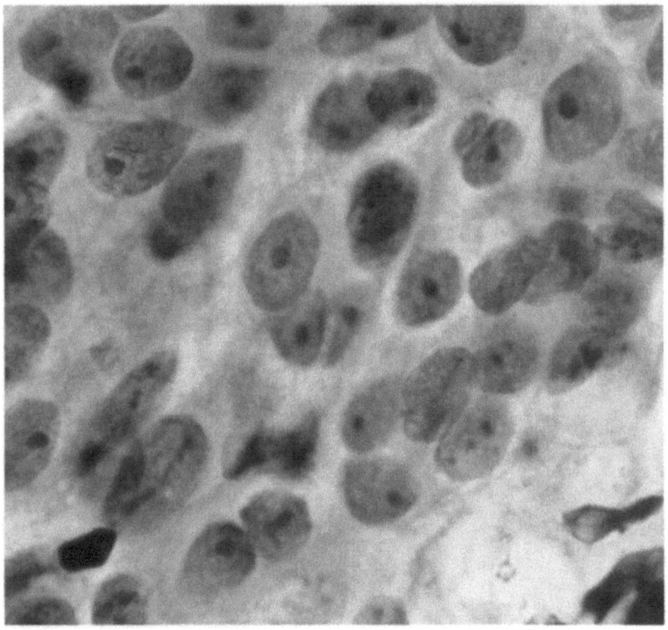

Fig. 89b. Same case at higher magnification. Despite the considerable size of the nuclei, the average nucleolar size and nucleolar variability are only medium grade. ×630

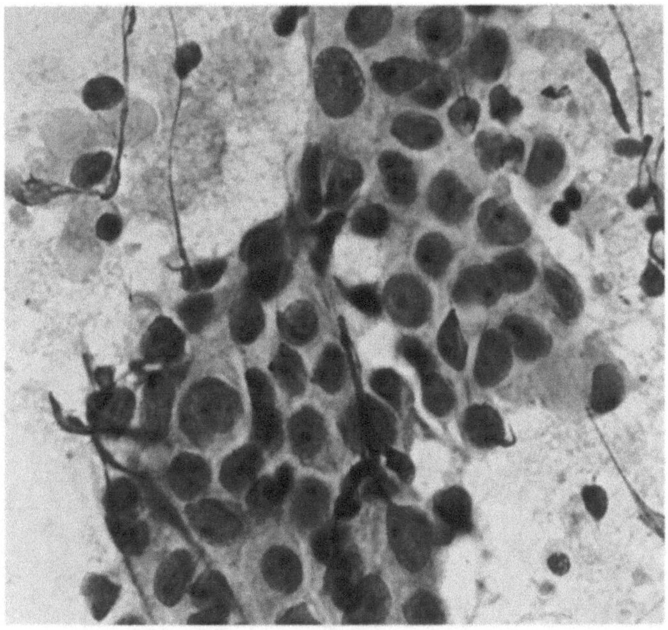

Fig. 90a. Grade II carcinoma: average nuclear size 3; variability in nuclear size 2; average nucleolar size 1; nucleolar variability 2; disturbance of nuclear arrangement 3; nuclear dissociation 2. Score: 13. ×400

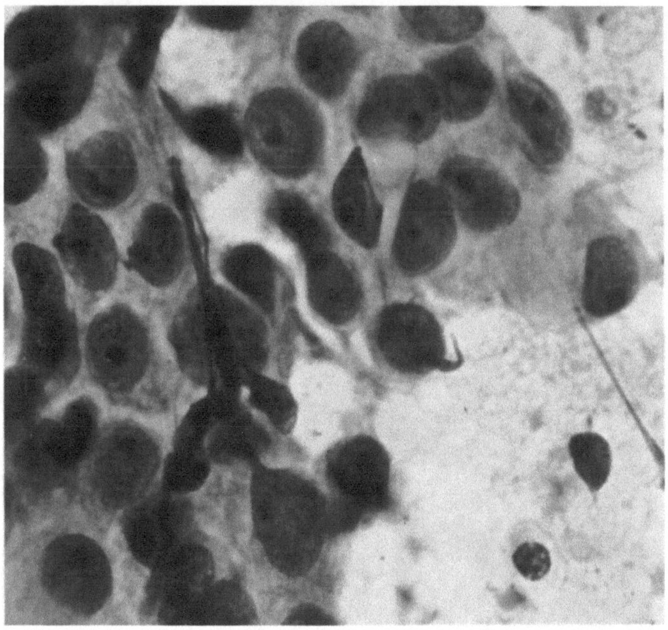

Fig. 90b. Same case at higher magnification. ×630

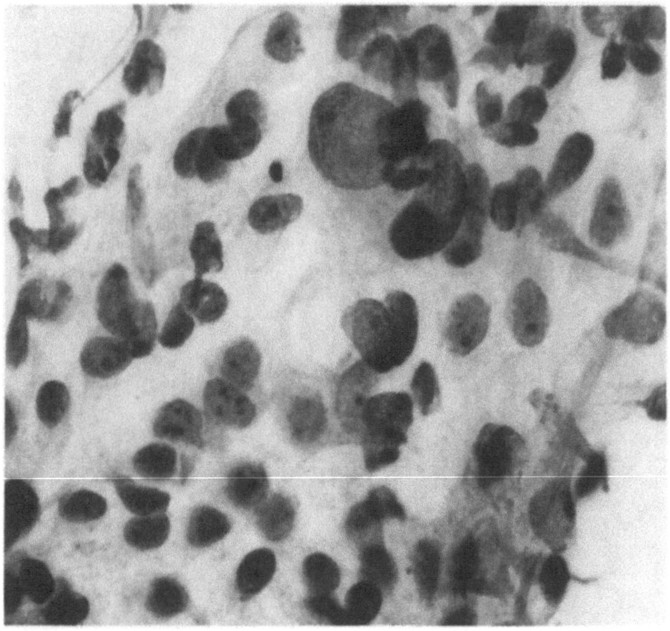

Fig. 91. Grade II carcinoma: average nuclear size 2; variability in nuclear size 3; average nucleolar size 1; nucleolar variability 2; disturbance of nuclear arrangement 3; nuclear dissociation 2. Score: 13. ×400

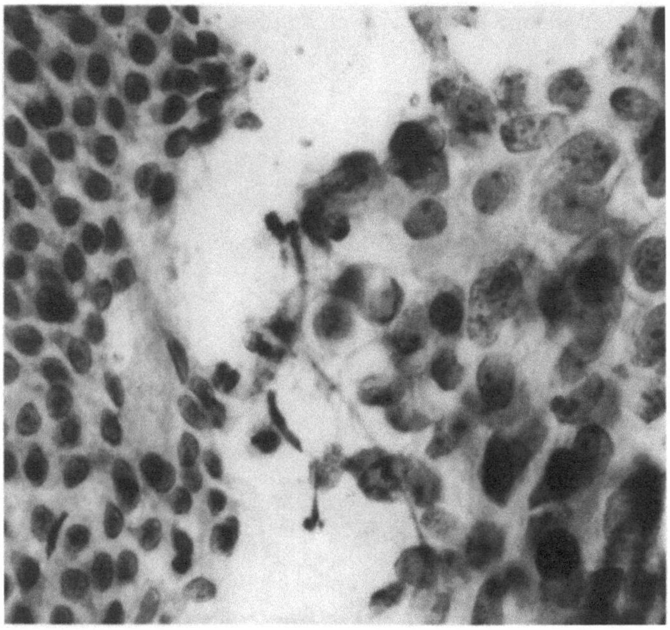

Fig. 92. Grade II carcinoma (*right*) and a sheet of rectal mucosal cells (*left*): average nuclear size 2; variability in nuclear size 2; average nucleolar size 1; nucleolar variability 3; disturbance of nuclear arrangement 3; nuclear dissociation 2. Score: 13. ×400

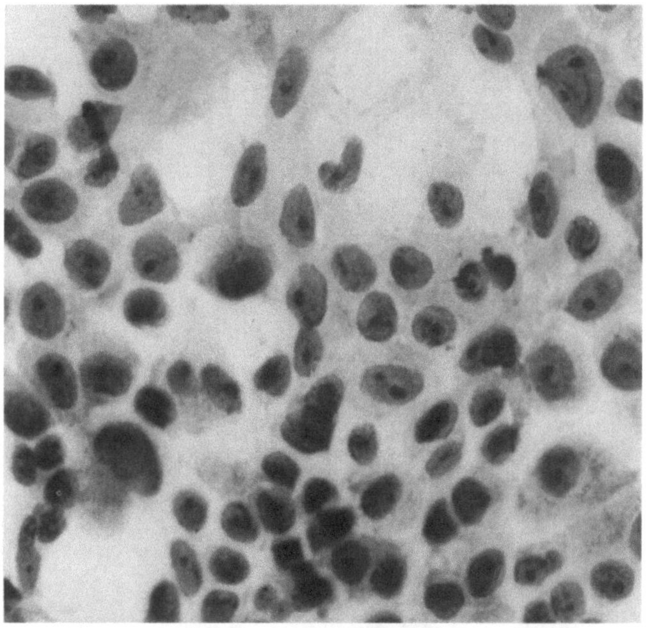

Fig. 93a. Grade II carcinoma: average nuclear size 2; variability in nuclear size 2; average nucleolar size 2; nucleolar variability 2; disturbance of nuclear arrangement 2; nuclear dissociation 2. Score: 12. ×400

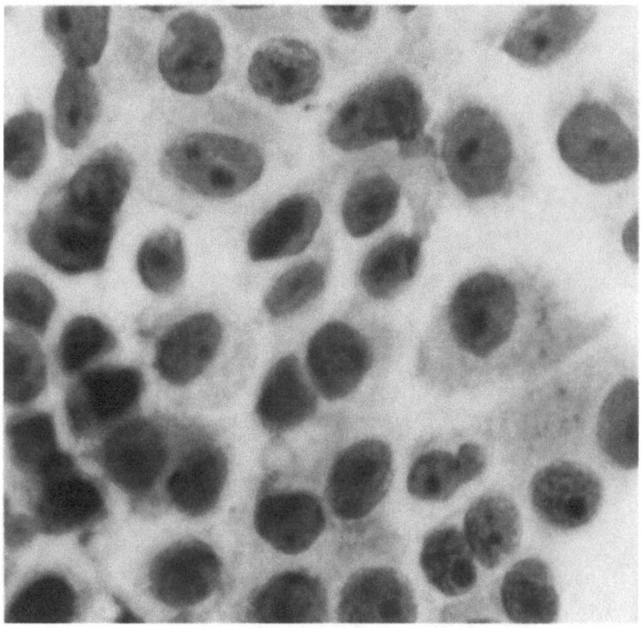

Fig. 93b. Same case at higher magnification: particularly clear representation of the parameter 'nucleolar variability'. ×630

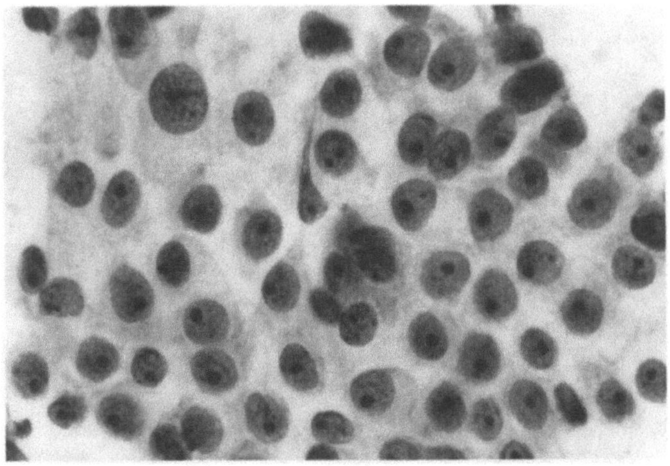

Fig. 94a. Grade II carcinoma: average nuclear size 2; variability in nuclear size 2; average nucleolar size 2; nucleolar variability 2; disturbance of nuclear arrangement 2; nuclear dissociation 1. Score: 11. ×400

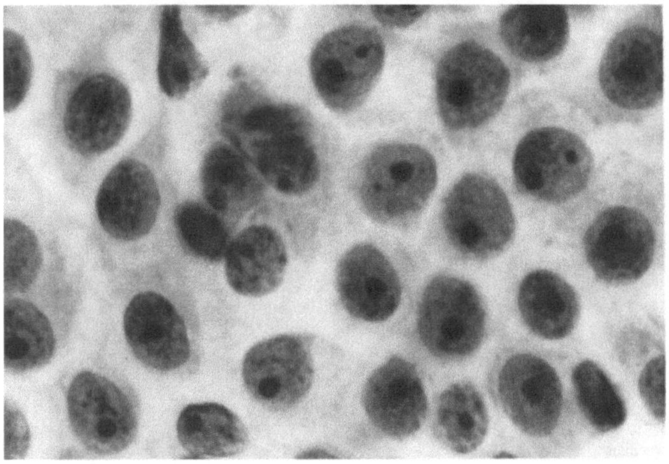

Fig. 94b. Same case at higher magnification: the parameter 'nucleolar variability' only becomes optimally evaluable at this magnification. ×630

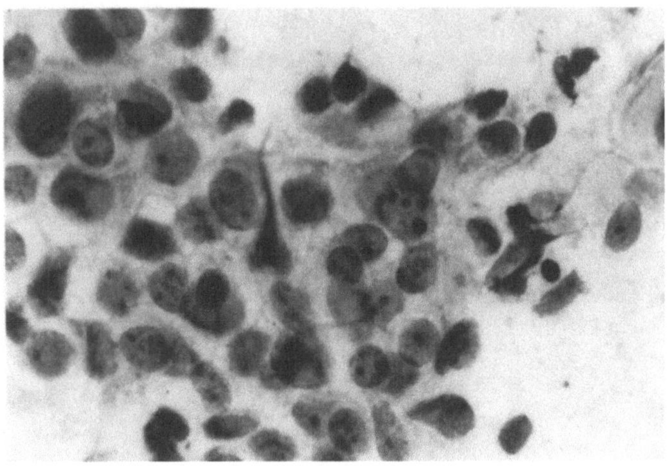

Fig. 95. Grade II carcinoma: average nuclear size 2; variability in nuclear size 3; average nucleolar size 2; nucleolar variability 3; disturbance of nuclear arrangement 2; nuclear dissociation 2. Score: 14. ×400

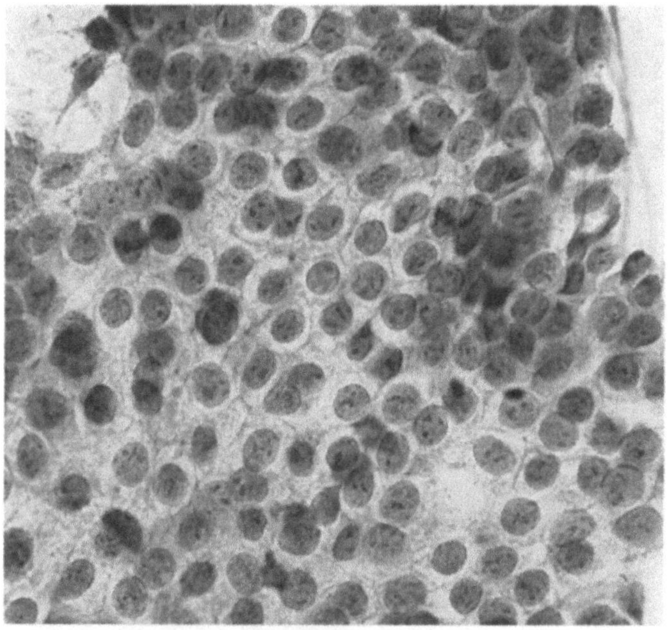

Fig. 96. Grade II carcinoma: average nuclear size 2; variability in nuclear size 1; average nucleolar size 1; nucleolar variability 3; disturbance of nuclear arrangement 2; nuclear dissociation 2. Score: 11. ×400

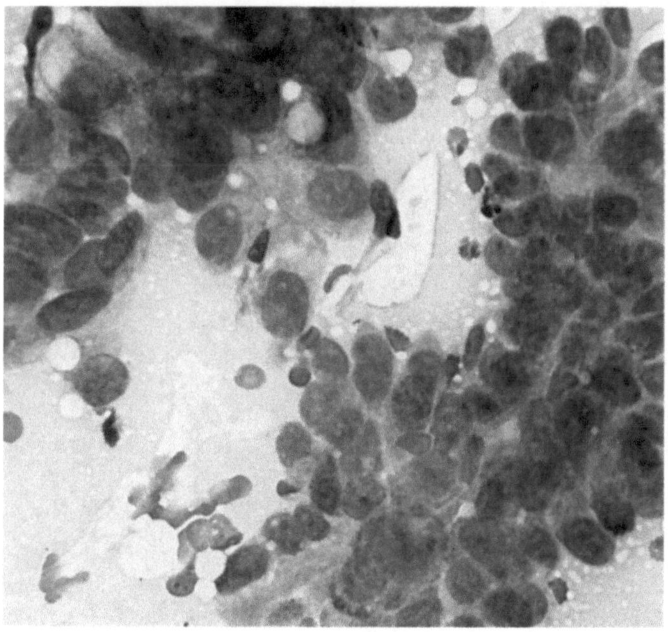

Fig. 97a. Grade II carcinoma, stained according to May-Grünwald-Giemsa. ×400

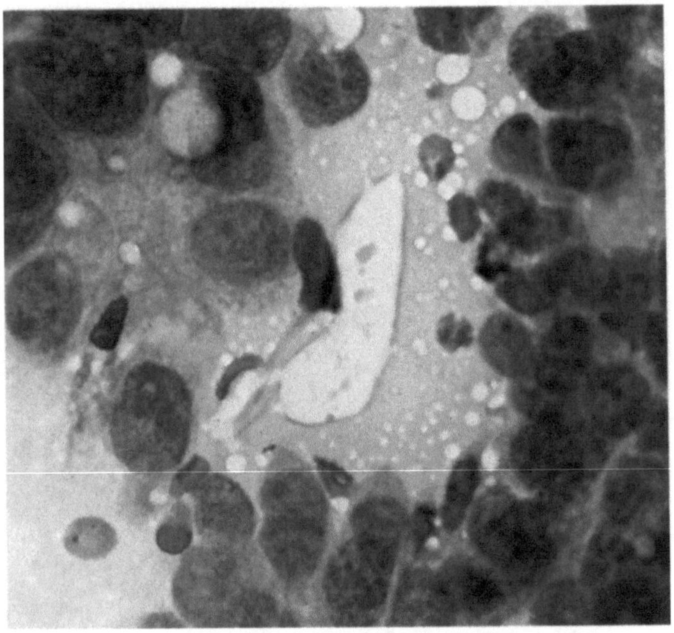

Fig. 97b. Same case at higher magnification. Figures 97a and b impressively demonstrate that at the same grade of malignancy the nuclei appear significantly larger and more polymorphic after MGG staining than after Papanicolaou staining. ×630

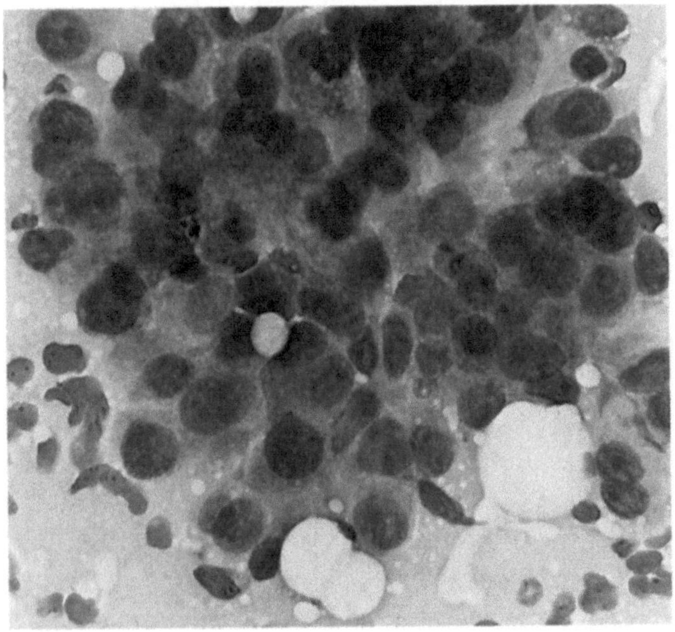

Fig. 98. Grade II carcinoma. MGG staining. ×400

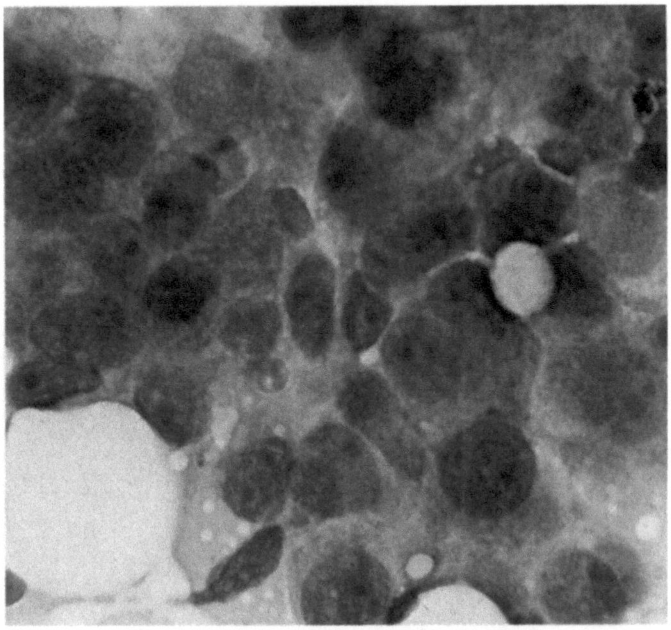

Fig. 99. Grade II carcinoma. MGG staining. ×630

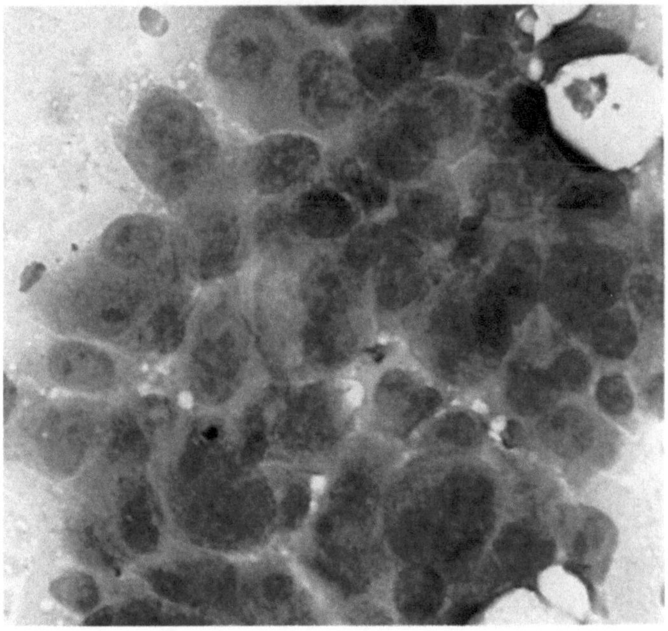

Fig. 100a. Grade II carcinoma. MGG staining. ×630

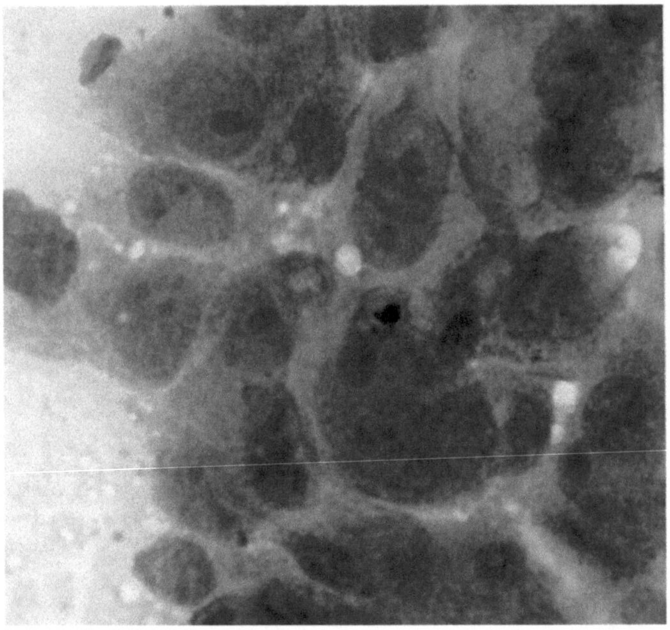

Fig. 100b. Same case at higher magnification. Oil immersion, ×1000

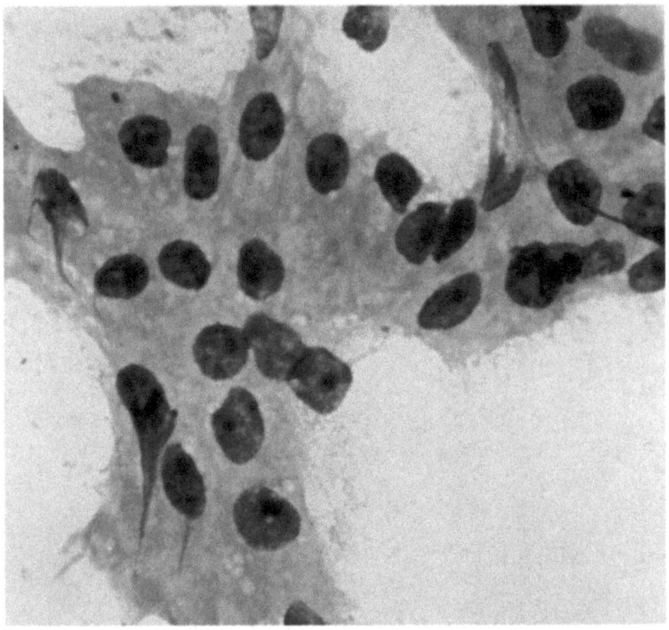

Fig. 101. Grade II carcinoma. MGG staining. ×400

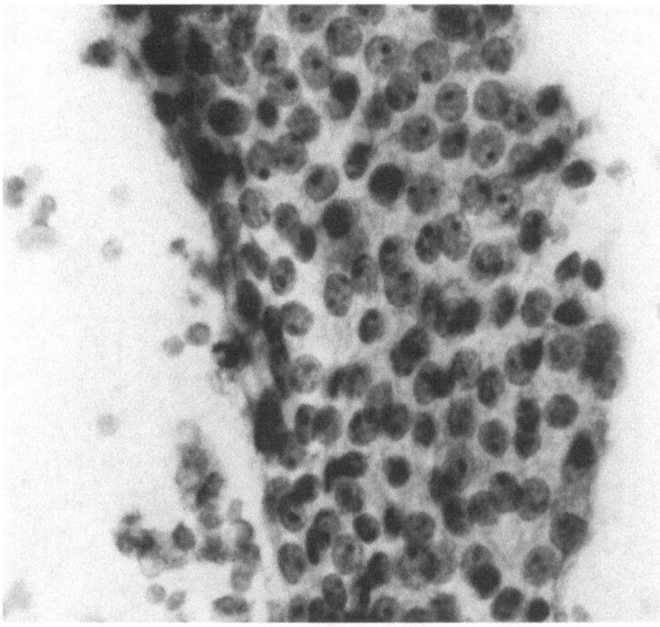

Fig. 102. Grade II carcinoma. Pap staining. Same case as in Fig. 98. After Pap staining the nuclei appear distinctly smaller than after MGG staining. ×400

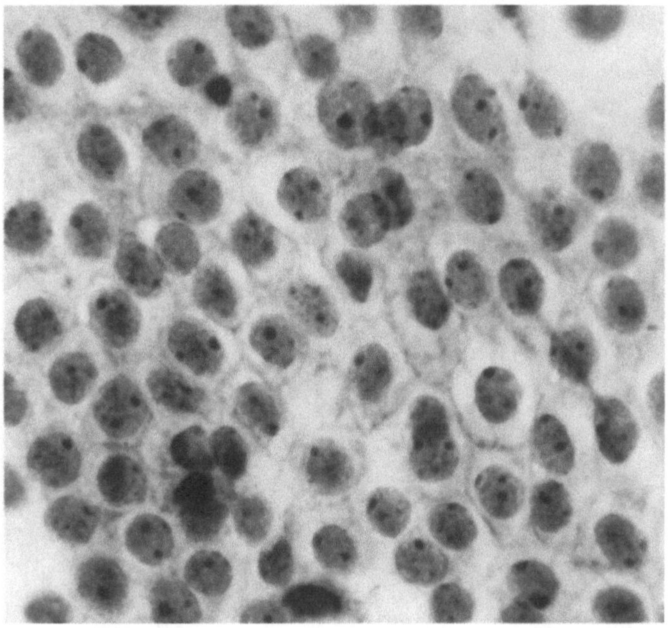

Fig. 103. Grade II carcinoma. Pap staining. Same case as in Fig. 101. The nuclei are here smaller than after MGG staining, and the nucleolar variability can be evaluated considerably better. ×400

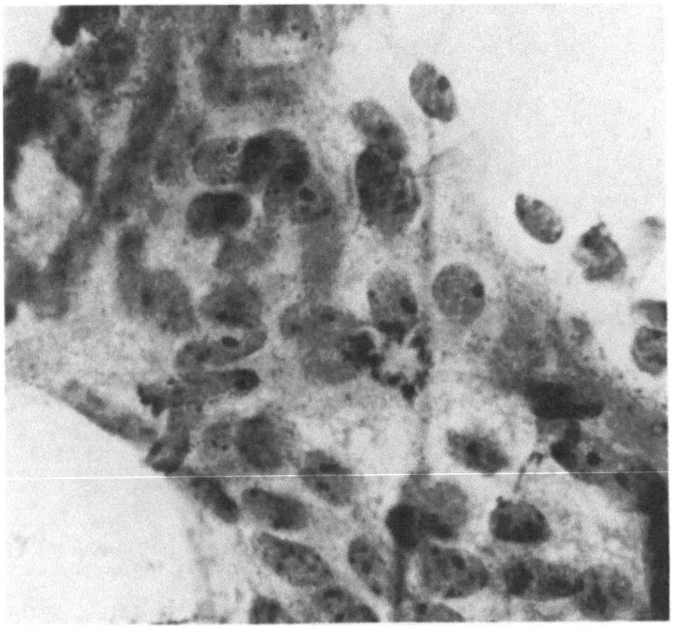

Fig. 104. Grade III carcinoma with the following evaluation of the six parameters decisive for grading: average nuclear size 3; variability in nuclear size 3; average nucleolar size 2; nucleolar variability 3; disturbance of nuclear arrangement 3; nuclear dissociation 2. Score: 16. ×400

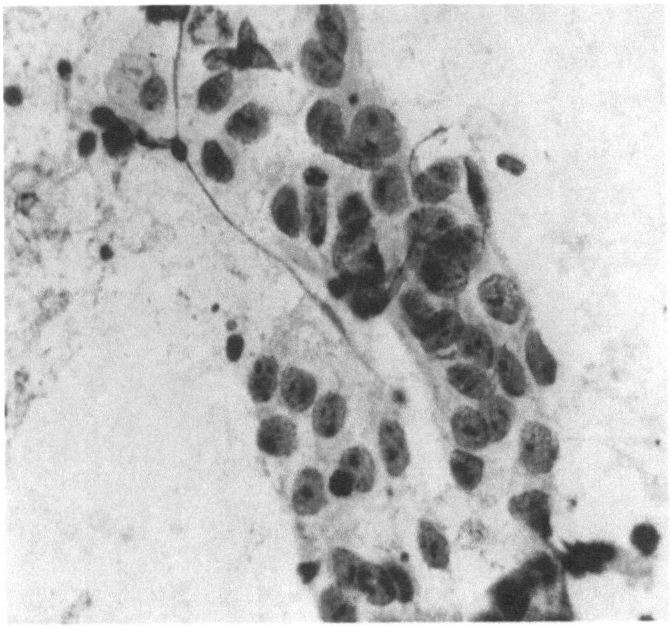

Fig. 105a. Grade III carcinoma: average nuclear size 2; variability in nuclear size 3; average nucleolar size 2; nucleolar variability 3; disturbance of nuclear arrangement 3; nuclear dissociation 2. Score: 15. ×400

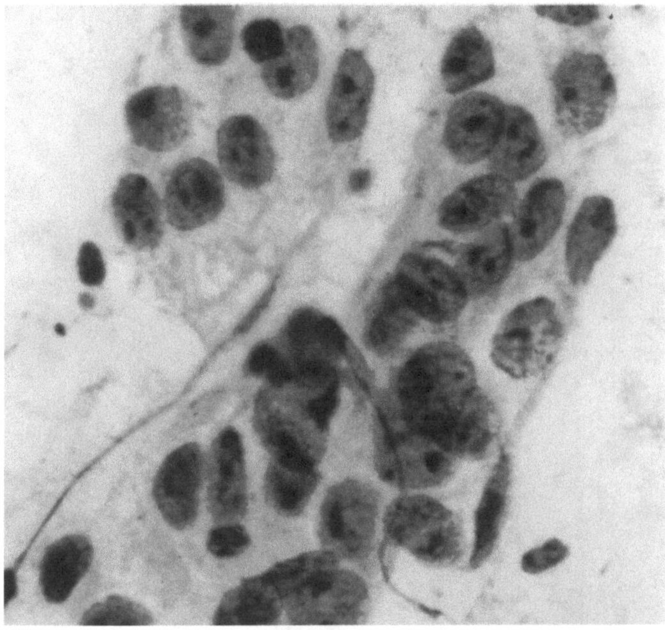

Fig. 105b. Same case at higher magnification. ×630

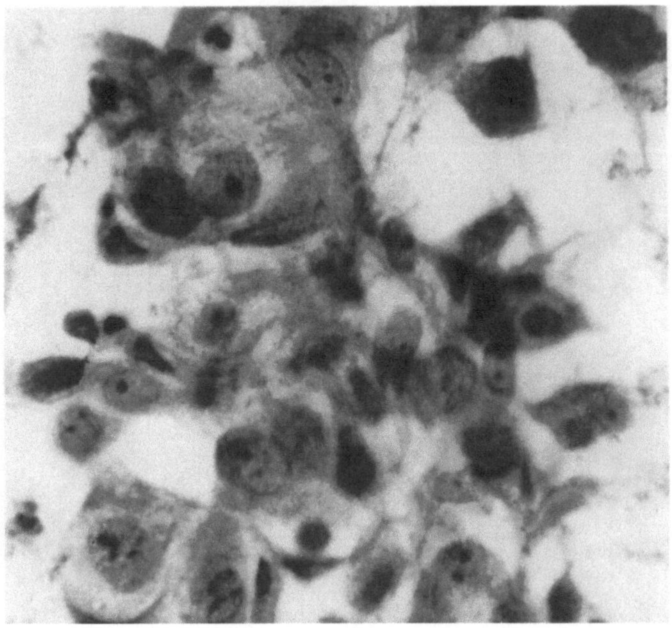

Fig. 106. Grade III carcinoma: average nuclear size 3; variability in nuclear size 3; average nucleolar size 3; nucleolar variability 2; disturbance of nuclear arrangement 3; nuclear dissociation 2. Score: 16. ×400

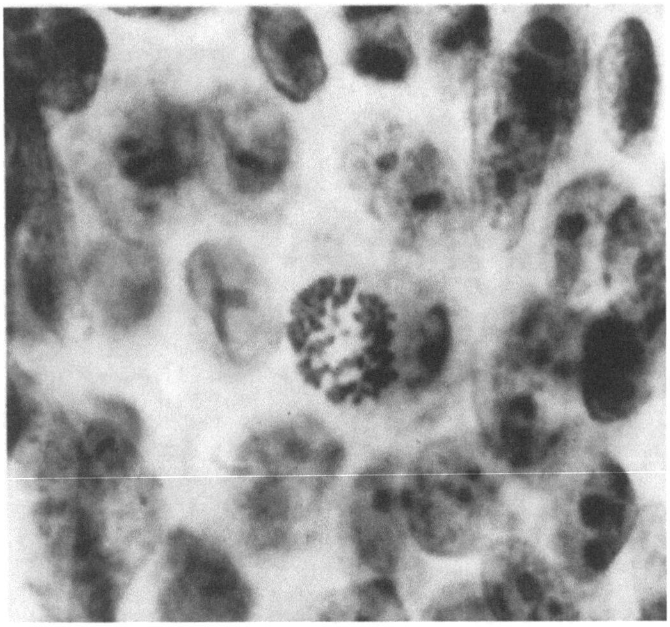

Fig. 107. Grade III carcinoma: all six parameters received the evaluation '3'. A cell undergoing mitosis is visible in the center. ×630

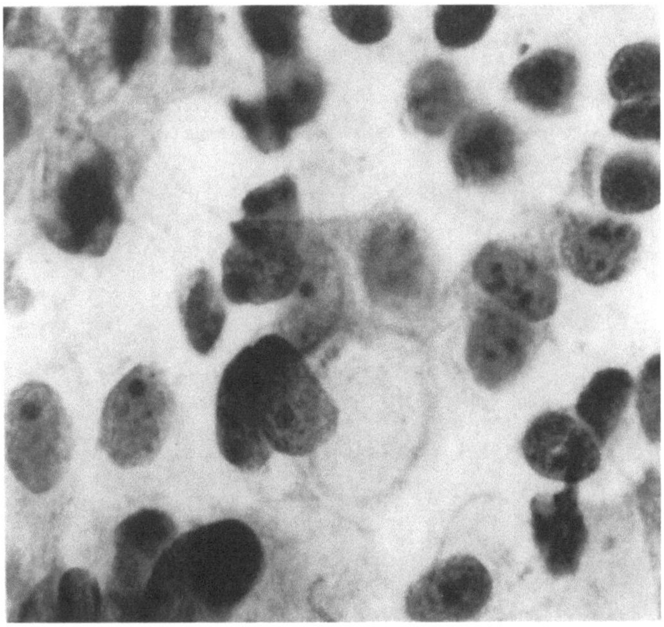

Fig. 108. Grade III carcinoma: average nuclear size 3; variability in nuclear size 3; average nucleolar size 3; nucleolar variability 2; disturbance of nuclear arrangement 2; nuclear dissociation 2. Score: 15. ×400

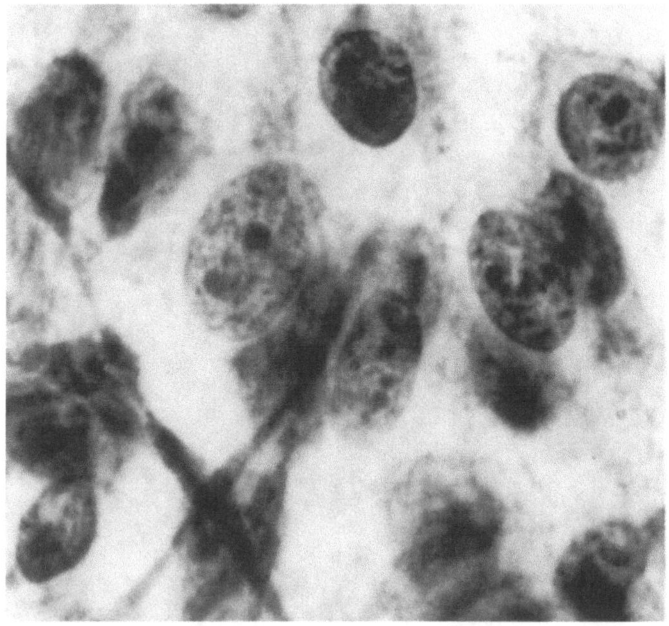

Fig. 109. Grade III carcinoma: All six parameters evaluated as '3'. In addition there is striking coarse clumping of chromatin. ×630

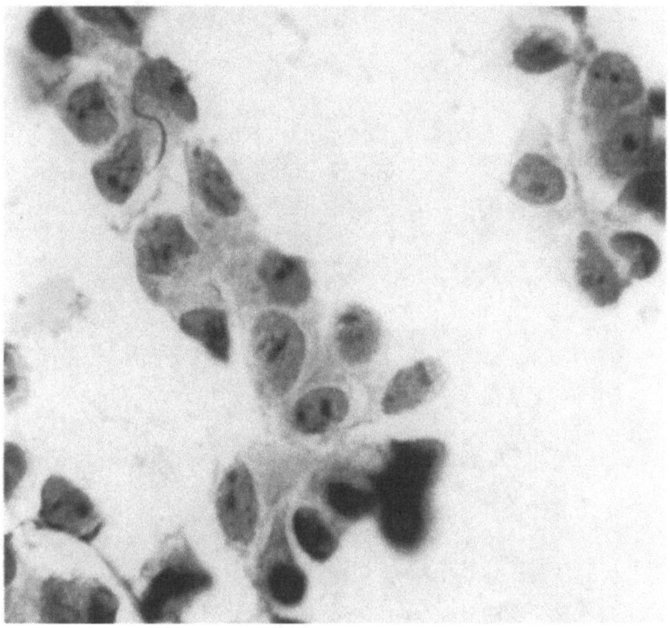

Fig. 110. Grade III carcinoma: average nuclear size 3; variability in nuclear size 3; average nucleolar size 2; nucleolar variability 3; disturbance of nuclear arrangement 2; nuclear dissociation 2. Score: 15. ×400

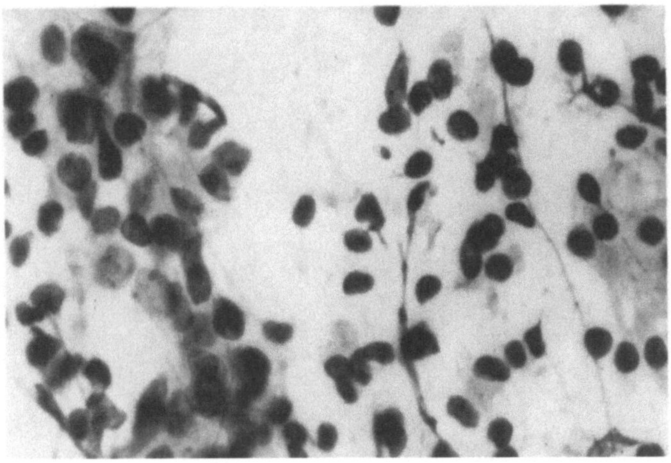

Fig. 111. Grade III carcinoma with very extensive nuclear dissociation. ×100

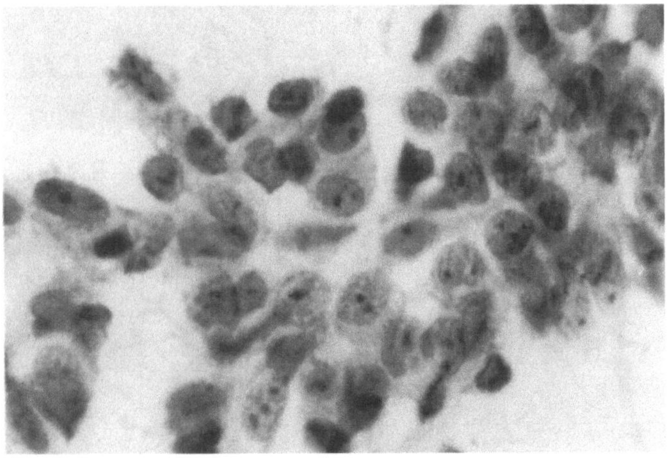

Fig. 112. Grade III carcinoma: average nuclear size 3; variability in nuclear size 3; average nucleolar size 2; nucleolar variability 2; disturbance of nuclear arrangement 3; nuclear dissociation 2. Score: 15. ×400

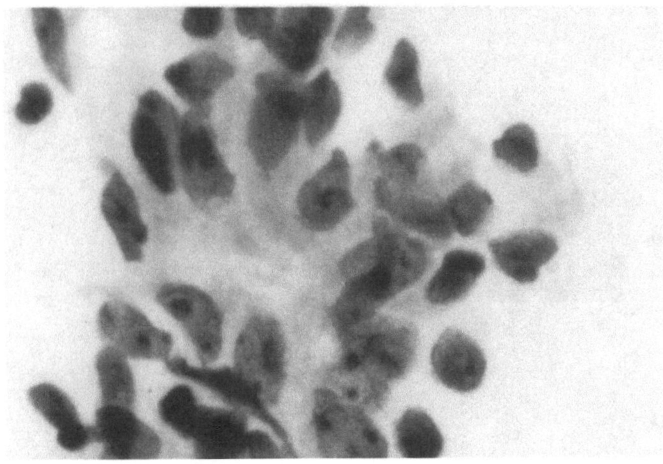

Fig. 113. Another group of cells from the grade III carcinoma shown in Fig. 112. ×400

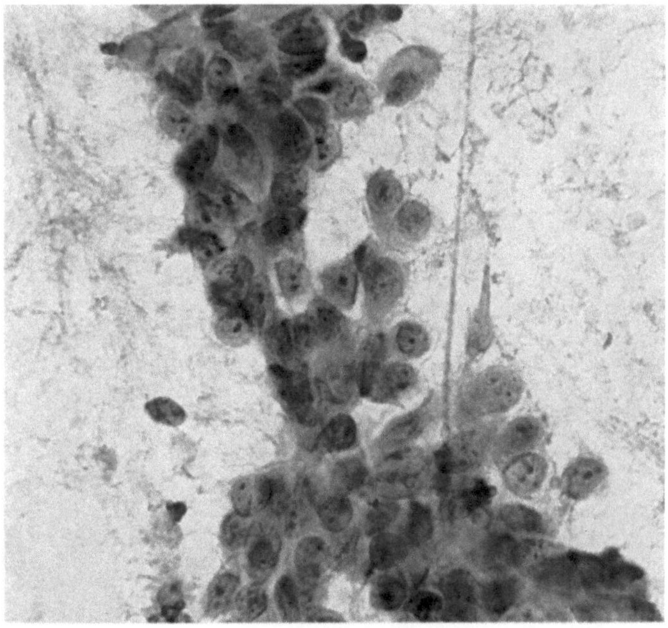

Fig. 114. Grade III carcinoma: average nuclear size 2; variability in nuclear size 2; average nucleolar size 2; nucleolar variability 3; disturbance of nuclear arrangement 3; nuclear dissociation 2. Score: 14. ×400

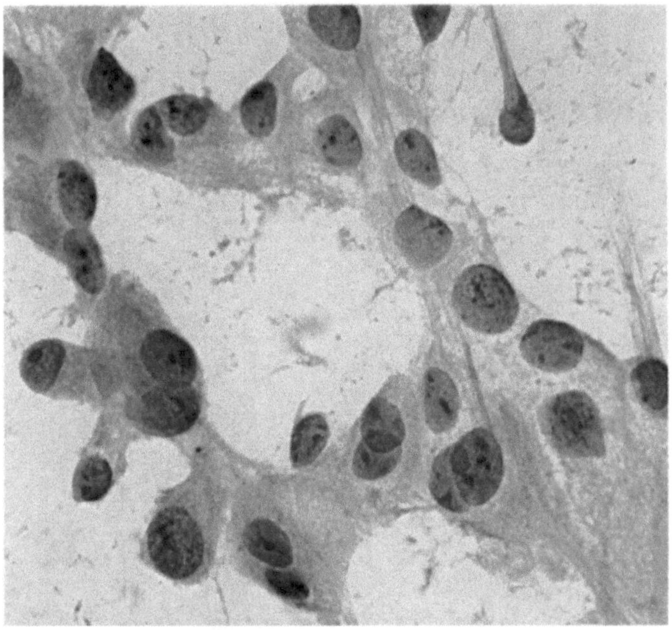

Fig. 115. Grade III carcinoma: average nuclear size 3; variability in nuclear size 3; average nucleolar size 2; nucleolar variability 3; disturbance of nuclear arrangement 2; nuclear dissociation 2. Score: 15. ×400

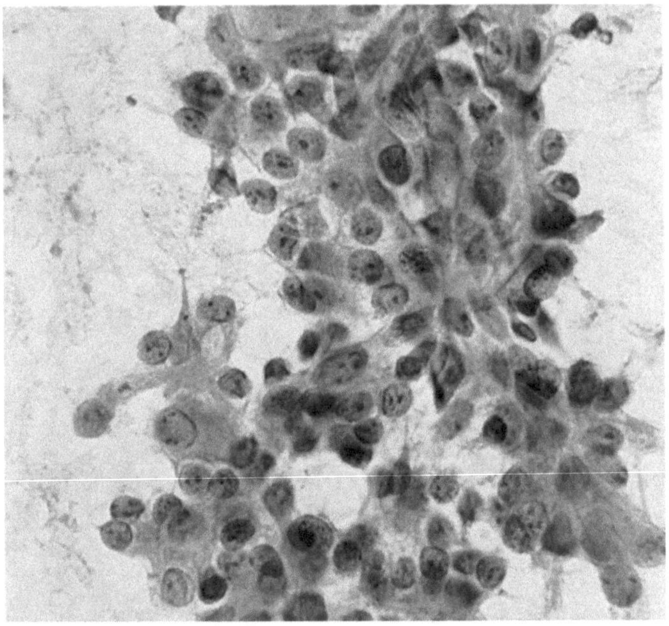

Fig. 116a. Grade III carcinoma at low magnification. Even at this magnification, the characteristics of malignancy are strikingly pronounced. ×100

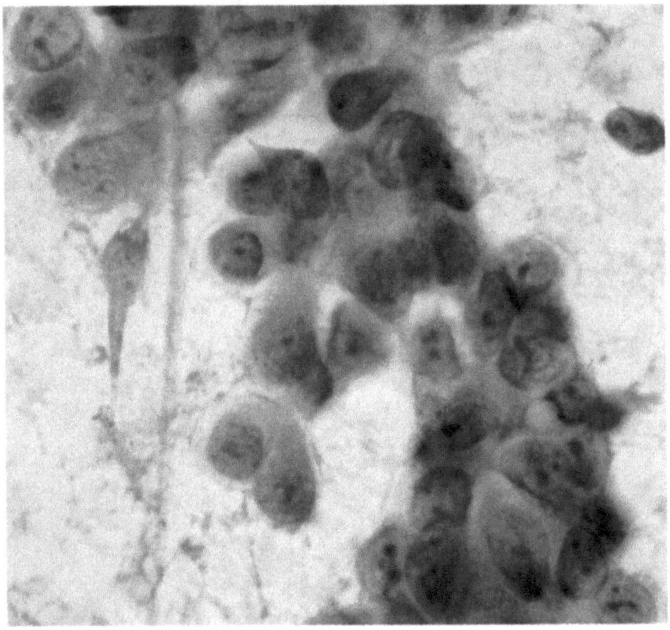

Fig. 116b. Same case, another group of cells at medium magnification: average nuclear size 3; variability in nuclear size 3; average nucleolar size 2; nucleolar variability 3; disturbance of nuclear arrangement 3; nuclear dissociation 3. Score: 17. ×400

9 Treatment Control by Means of Regression Grading

Histological or cytological examination of a biopsy specimen from the primary tumor is at present the only objective and reproducible procedure for determining the effect of treatment on the primary tumor in cases of locally advanced, inoperable prostatic carcinoma (COSGROVE et al. 1973; SEWELL et al. 1975). At the same time, biopsies performed during the course of treatment constitute a unique method of obtaining information on the biological activity of the tumor, i.e. its neoplastic potency, above all when there are no signs that a tumor has reached a far advanced stage (hydronephrosis, metastases) clinically.

Locally advanced, asymptomatic prostatic carcinoma (stage $T_3/T_4N_0M_0$) is virtually the only tumor whose progression can be slowed by various therapeutic modalities. In such patients, therefore, treatment – usually anti-androgen (castration, estrogens, anti-androgens) – is as a rule instituted immediately after the diagnosis has been secured cytologically or histologically, insofar as radiotherapy does not appear to be indicated.

Since it is known that some 20% of tumors do not respond to anti-androgen therapy, aspiration biopsy is of paramount importance in controlling the effect treatment is having on the primary tumor. In addition, it is an absolutely ideal method for this purpose, since the negligible stress placed upon the patient and the *minimal complication rate* mean that aspiration biopsy can be performed at regular intervals of approximately 3–6 months.

For these reasons aspiration biopsy is to date the only objective and tolerable procedure which can be used to demonstrate a satisfactory or good therapeutic response on the basis of appropriate signs of regression, or a lack of effect when no such signs are evident.

In particular, evidence that the treatment is having only an inadequate effect or none at all is the *earliest indication of imminent progression,* and usually arises months *prior to* the first signs of clinically progressing tumoral disease (urinary retention, metastases). This affords the possibility of switching to a secondary or tertiary treatment *before* progression is clinically evident (LEISTENSCHNEIDER and NAGEL 1980).

Signs of regression induced by therapy can be assessed both *histologically* and *cytologically,* the two methods being of equal value.

The *histological signs of regression* are identified by evaluation of glandular structure, stroma and alterations in the tumor cells themselves (SCHENKEN et al. 1942; ALKEN et al. 1975; SPIELER et al. 1976; KASTENDIECK et al. 1980).

The *cytological signs of regression* consist in the nuclear and cytoplasmic changes demonstrated during the treatment (ESPOSTI 1971; FAUL 1975; SPIELER et al. 1976; LEISTENSCHNEIDER and NAGEL 1980).

Neither the histological nor the cytological signs of regression in treated prostatic carcinoma are therapy specific; rather they apply equally to all therapeutic modalities used today!

In the past, *cytological treatment control* has above all been employed after:

- *estrogen treatment* (Esposti 1971; Faul 1975; Rost et al. 1976; Keller et al. 1981), and
- *local irradiation* (Spieler et al. 1976; Kurth et al. 1977).

On the basis of 600 aspiration biopsies performed in 260 patients, we ourselves have investigated the correlation between the cytological regression grade and the clinical course for six different types of treatment, undertaken sequentially in some patients **(Table 14).**

Insufficient cellular material was aspirated in an average of 5.2% of cases (range for the six different treatments: 2.5–12%), however, no connection was observed between form of treatment and unsatisfactory material **(Table 15).**

Overall the percentage of aspirates that could not be evaluated was similar to the rates obtained at other centers with experience in primary diagnosis **(Table 1,** p. 3).

Table 14. Overview of the various forms of treatment, the number of patients in whom they were employed and the number of control aspiration biopsies performed (Leistenschneider and Nagel 1983)

Treatment	*n*	Biopsies
Estrogens ± bilat. orchiectomy	138 (53.1%)	312 (52%)
Anti-androgen	4 (1.5%)	15 (2.5%)
Estracyt/Emcyt	61 (23.5%)	120 (20%)
Irradiation	31 (11.9%)	105 (17.5%)
Irradiation + hormone therapy	12 (4.6%)	25 (4.2%)
Cyclophosphamide/5-FU	14 (5.4%)	23 (3.8%)
	260 (100%)	600 (100%)

Table 15. Frequency of inadequate cellular material in control aspiration biopsies, in relation to the various forms of treatment

Treatment	Biopsies	Inadequate cellular material
Estrogens ± bilat. orchiectomy	312	18 (5.8%)
Anti-androgen	15	1 (6.7%)
Estracyt/Emcyt	120	3 (2.5%)
Irradiation	105	4 (3.8%)
Irradiation + hormone therapy	25	3 (12.0%)
Cyclophosphamide/5-FU	23	2 (8.7%)
	600	31 (5.2%)

9.1 Cytological Signs of Regression

The cytological characteristics of regression can be optimally assessed in wet-fixed smears stained with Papanicolaou's stain.

Cytologically, signs of regression can be demonstrated both in the *nucleus* and in the *cytoplasm* **(Table 16).**

In addition to the decisive signs of regression in the nucleus, *megakaryosis* (Spieler et al. 1976; Leistenschneider and Nagel 1980) is also encountered, especially in irradiated patients. *Nuclear vacuolation* is infrequent, occurring in some 10% of cases only.

The *squamous cell metaplasia* which can often be demonstrated alongside the above phenomena **(Figs. 123, 124)** is *not* a genuine characteristic of regression. Rather it is merely indicative of a possible effect of estrogen upon the prostate, though it is not even specific for this (Spieler et al. 1976; Leistenschneider and Nagel 1983).

Metaplastic squamous cells are predominantly observed in treated grade I or grade II carcinomas, i.e. they are generally found in

Table 16. Cytological treatment control of prostatic carcinoma: signs of regression in the nucleus and cytoplasm

	Signs of regression
Nucleus	Reduction in size Pyknosis Nucleolar size reduced Nucleolus indistinguishable Loose chromatin pattern Regular distribution of chromatin Nuclear membrane distinct Vacuolation Megakaryosis
Cytoplasm	Vacuolation Shrinkage 'Lake' formation

Table 17. Classification of the cytologically detectable signs of regression

Signs of marked regression	Pyknosis Nucleoli reduced in size or not detectable Loose chromatin pattern Regular chromatin distribution Nuclear membrane distinct Reduction in nuclear size Rarefaction of nuclei Cytoplasmic lakes
Signs of slight regression	Nuclear vacuoles Cytoplasmic vacuoles Cytoplasmic shrinkage

cases in which the therapeutic effect has been good.

Two groups of therapy-induced regression signs are distinguished on the basis of their nature and degree **(Table 17):**

signs of marked regression
signs of slight regression

9.1.1 Signs of Marked Regression (Table 17)

The following characteristics are *always* present when marked regression has occurred, irrespective of the tumoral stage, the grade of differentiation or the type of treatment:

- reduction in nuclear size **(Figs. 117b, 118b)**
- pyknosis **(Figs. 118b, 118c)**
- diminution or disappearance of nucleoli **(Figs. 117b, 118b, 119)**
- loosened or regularly distributed chromatin **(Figs. 118b, 119, 120a)**
- clearly recognizable nuclear membranes **(Figs. 118b, 119, 120a)**
- rarefaction **(Figs. 117b, 118c, 119, 121a)**
- cytoplasmic 'lakes' as a consequence of pronounced nuclear rarefaction **(Figs. 119, 138a).**

When pronounced regression has occurred, it is also always possible to identify aggregates of phagocytizing *histiocytes,* some of which have several nuclei **(Figs. 127b–131).**

9.1.2 Signs of Slight Regression (Table 17)

It is characteristic of slight regressive changes that the following characteristics are absent in the vast majority of the groups of cells:

- reduction in nuclear size **(Figs. 145b, 146b)**
- diminution or disappearance of nucleoli **(Figs. 145b, 146b, 148).**

Nevertheless, the following signs of regression are present:

- cytoplasmic vacuoles **(Figs. 122, 148, 154)**
- shrinkage of the cytoplasm **(Fig. 146c)**
- nuclear vacuolation **(Fig. 155a).**

9.2 Cytological Regression Grading (Table 18)

Cytological regression grading was developed with close reference to histological regression grading (Alken et al. 1975). It forms the basis for the cytological assessment of the effect of treatment on cellular material from the primary tumor: six grades of regression are distinguished (0, II, IV, VI, VIII and X), whereby the therapeutic effect ranges from 'very good' through 'satisfactory' to 'none' **(Table 18).**

Table 18. Cytological regression grading

Regression grade	Cytomorphological findings	Therapeutic effect
0	Normal sheets of cells, isolated low-grade atypias (Pap II). No tumor cells, many macrophages	Very good
II	Moderate atypias (maximum Pap III). Many normal sheets of cells. Conclusive classification as carcinomatous no longer possible. Many macrophages	Good
IV	Small isolated groups of carcinoma cells. Signs of marked regression. Many normal sheets of cells ('residual carcinoma')	Satisfactory
VI	Many groups of carcinoma cells. Signs of marked regression	Moderate
VIII	Many groups of carcinoma cells. Signs of slight regression	Poor
X	Groups of carcinoma cells without signs of regression	No effect

9.2.1 Cytomorphological Criteria of Regression Grading

9.2.1.1 Regression Grade 0

Alongside unremarkable prostatic epithelium, only slightly atypical sheets of prostatic epithelial cells (Pap II) are detectable. In many sheets regressive changes occur especially in the cytoplasm, in the form of vacuolation and shrinkage. Numerous squamous cells – isolated or in small groups scattered over the entire smear – are frequently seen (as in regression grade II) **(Figs. 132b, 132c).**

Phagocytizing histiocytes (macrophages) having several or many nuclei and a foamy granular content are strikingly frequent, as are phagocytized nuclear debris and squamous cells **(Figs. 127b–131).** It is not unusual for the cytoplasm of these macrophages to be permeated by particularly large vacuoles **(Figs. 128b, 128c).**

The cytological diagnosis of regression grade 0 can be made reliably only when at least 20 medium-sized or large sheets of cells from each prostatic lobe are available for scrutiny!

9.2.1.2 Regression Grade II

In addition to a preponderance of unremarkable prostatic epithelium, isolated groups of cells display moderate atypia (Pap III), mainly in the form of condensation of the nuclear chromatin, moderate variations in nuclear size and indistinct cell borders **(Fig. 132b).** It is *not* possible to classify such groups as carcinomatous on the basis of cytology. Numerous squamous cells and macrophages, alone or in groups distributed over the smear, occur very frequently in regression grade II also **(Figs. 132b, 132c).**

9.2.1.3 Regression Grade IV

The smear exhibits mainly unremarkable prostatic epithelium. In addition, however, small and medium-sized groups of cells containing a few typical carcinoma cell nuclei are demonstrated; apart from the density and irregular distribution of the chromatin, these nuclei occasionally contain prominent nucleoli. The nuclear arrangement is usually disturbed **(Figs. 133b, 133c).**

Quantitatively, however, these characteristics are of minor significance in comparison with the only moderately atypical nuclei containing at most small or conspicuous (but never prominent) nucleoli which are found in the same group of cells. The *nuclear membranes* are generally clearly visible **(Figs. 133c, 135).** In contrast to the findings *prior to* treatment **(Fig. 133a),** the marked *rarefaction of the nucleus* is particularly noticeable **(Figs. 133c, 134b, 135).** The *chromatin* is predominantly loosely granular in structure, and the cytoplasm clearly loosened or vacuolated. Furthermore, such smears always contain mostly normal or only slightly atypical groups of cells. As also in regression grades 0 and II, in regression grade IV *macrophages* are a regular feature **(Fig. 136).** Formation of cytoplasmic lakes is also typical.

9.2.1.4 Regression Grade VI

Apart from a few unremarkable sheets of prostatic epithelium, primarily groups of carcinoma cells are found at this regression grade. Numerous nuclei in these groups still display important characteristics of carcinoma – in particular, *markedly condensed chromatin* and conspicuous, in part even *prominent, nucleoli.* In the same groups of cells other *nuclei* contain nucleoli that are either scarcely distinguishable or are at most conspicuous. In these nuclei the *chromatin* is frequently loosely structured and the *nuclear membranes* distinct **(Figs. 139b, 141b).** In addition, nuclear rarefaction is always present, though it is less pronounced than at regression grade IV.

Typical for regression grade VI, then, is the presence of groups of carcinoma cells which clearly outnumber the sheets of normal epithelial cells **(Figs. 139b, 140b, 141b, 142b, c).**

On occasion, isolated groups of carcinoma cells without more pronounced signs of regression are also observed. Although the overall classification can still be regression grade VI if there are fewer such groups than those representing grade VI, the supplementary remark 'isolated regression grade VIII' should be added since such patients must be adjudged to have a worse prognosis than patients with 'pure' regression grade VI (see R VI–VIII, **Table 21,** p. 144).

9.2.1.5 Regression Grade VIII

The cytological smear is characterized by *numerous groups of carcinoma cells, the great majority of which show only signs of slight regression* – above all, vacuolation, shrinkage or loosening of the cytoplasm **(Figs. 145b, 146b, c, 147, 148, 154).** Mitoses are rather frequently seen **(Figs. 146c, 147).** While signs of marked regression may be observed in some of these groups, e.g. small nuclei, pyknosis and merely conspicuous nucleoli, these are quantitatively of minor significance compared with those groups in which scarcely any regressive changes are evident **(Figs. 149–152).**

The fact that in the same smear isolated groups of carcinoma cells may display marked regressive changes should not receive consideration in the definitive diagnosis, since their presence has no prognostic significance.

9.2.1.6 Regression Grade X

At this grade of regression, cytological smears are dominated by groups of carcinoma cells *without any* signs of regression. It predominantly occurs when there is resistance to various forms of treatment. It is characterized by very pronounced indications of malignancy, in particular by the presence of very large nuclei and great variability in nuclear size. The chromatin is extremely dense, often displaying marked clumping. The nucleoli are strikingly prominent and often polymorphic; their incidence per nucleus is increased **(Fig. 157a, b).** Cell borders are generally completely absent. The nuclear arrangement is irregular; there is pronounced dissociation of nuclei, and it is not infrequent for cell groups to show complete disintegration. The picture is one of polymorphic, naked tumor cell nuclei **(Fig. 156b, c).**

9.3 Reproducibility

The method of cytological regression grading described above forms the basis for assessing reproducibility.

In 70 consecutive personal investigations, *intraindividual reproducibility* averaged 90% over the six regression grades (83–100%) **(Table 19).** The variations were not statistically significant.

Only the varying classification between regression grades VI and VIII could have clinical significance, especially if a relatively good therapeutic result (R VI) were to be mistakenly interpreted as a poor effect (R VIII), for in such cases – particularly where the clinical picture is stable – no change of treatment would have been indicated had the grading been correct.

Table 19. Intraindividual reproducibility of cytological regression grading

Regression grade	*n*	Reproducible findings *n*	Reproducible findings %
0	2	2	100
II	6	5	83
IV	14	12	86
VI	26	23	88
VIII	18	17	94
X	4	4	100
	70	63	90

Table 20. Interindividual reproducibility of cytological regression grading

Regression grade	1st diagnostician *n*	2nd diagnostician (%)	Reproducibility
0	2	2	100
II	4	4	100
IV	16	13	81
VI	31	25	81
VIII	24	22	92
X	5	3	60
	82	69	84

The *interindividual reproducibility,* assessed by comparing the findings of two experienced diagnosticians on 82 consecutive aspiration biopsies, averaged 84% (60–100%) **(Table 20).** Again, the variations were not statistically significant. The apparently pronounced divergence in cases assigned to regression grades VIII and X would in any case have *no* clinical relevance even if statistically significant, since these grades are equally indicative of a poor therapeutic effect or even no effect at all.

The correlation between *histological and cytological regression grading* was 85.3% in our own observations on 102 cases (Leistenschneider and Nagel 1980).

There were six patients with cytological regression grade VIII in whom simultaneous punch biopsy resulted in the histological regression grade VI, which is associated with a more favorable prognosis. Repeat aspiration biopsy again indicated regression grade VIII, and within 1 year the clinical progression to be expected on the basis of the adverse *cytological* evaluation (R VIII) did in fact occur.

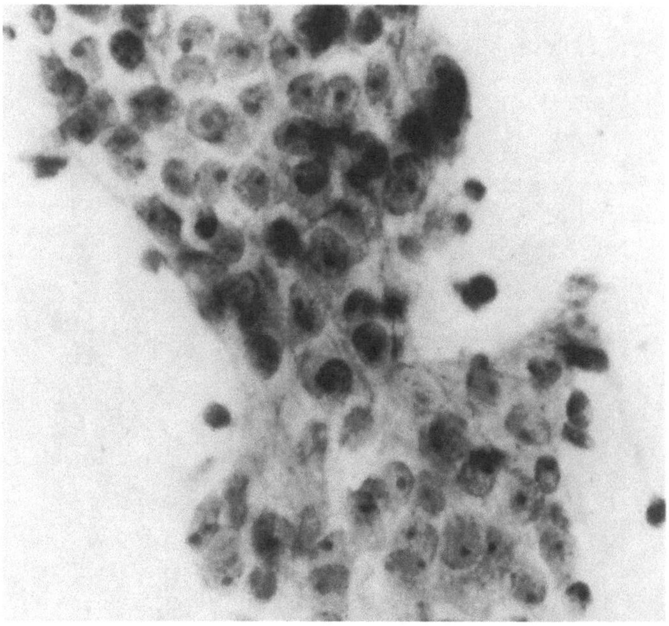

Fig. 117a. Grade II carcinoma prior to treatment. ×400

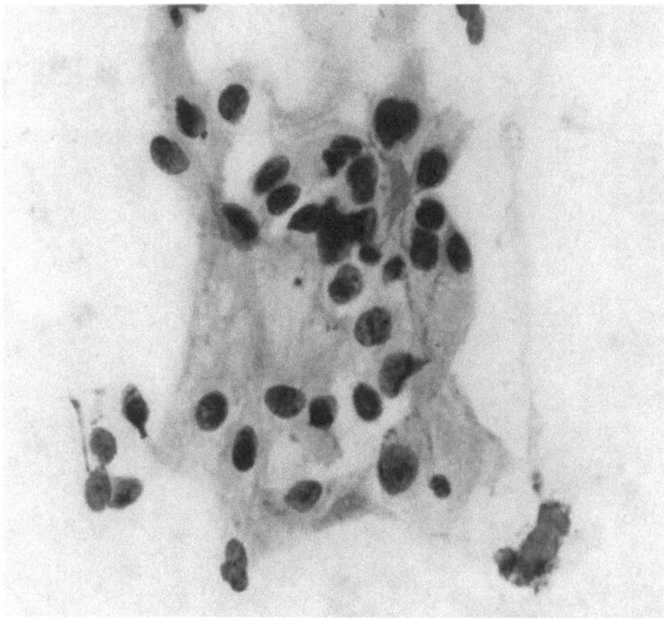

Fig. 117b. Same case after 3 months of hormone therapy: nuclei markedly smaller; rarified, considerably smaller nucleoli, some of which are indistinguishable; overall, signs of marked regression. ×400

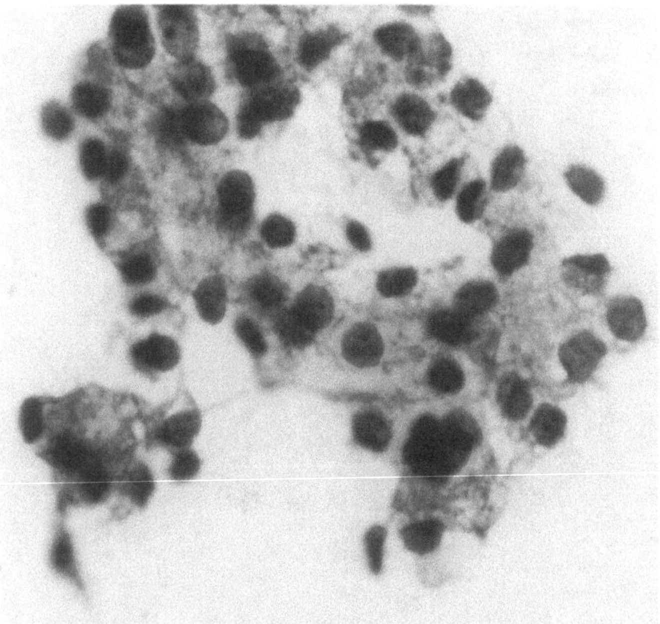

Fig. 117c. Same case, a different group of cells: pronounced fine vacuolation of the cytoplasm. Nuclear changes as in Fig. 117b. ×400

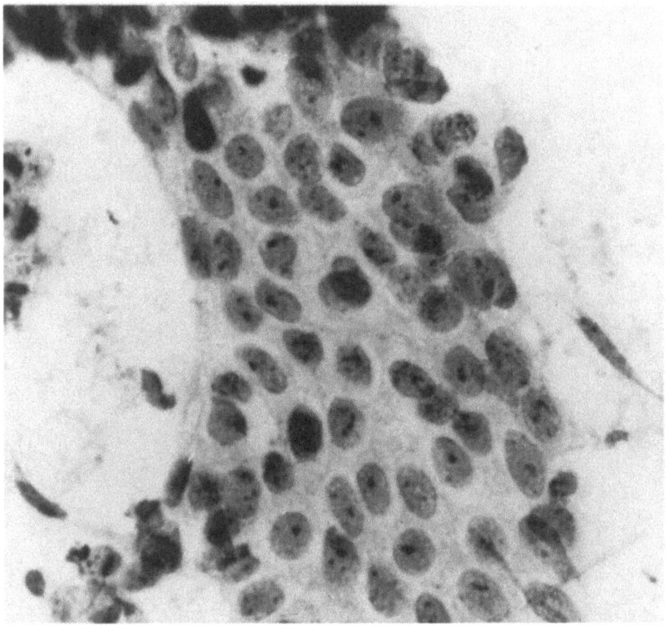

Fig. 118a. Grade II carcinoma prior to treatment. ×400

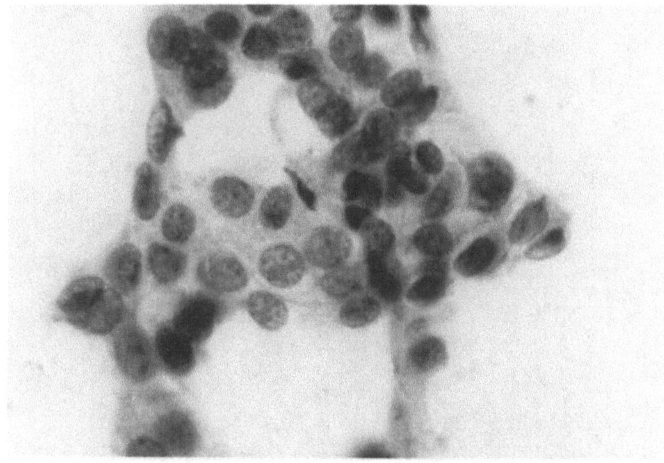

Fig. 118b. Same case after hormone therapy for 6 months. Signs of marked regression: reduction in nuclear size, isolated pyknotic nuclei, diminution or disappearance of the nucleoli. ×400

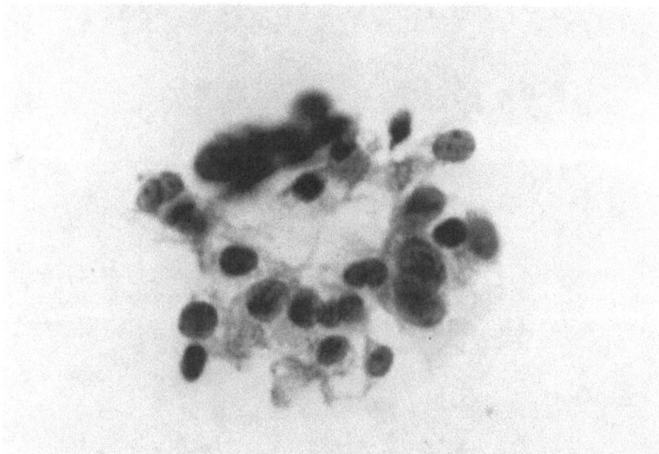

Fig. 118c. Same case, another group of cells. Several pyknotic nuclei and marked rarefaction of the nuclei in comparison with the cytoplasm present. ×400

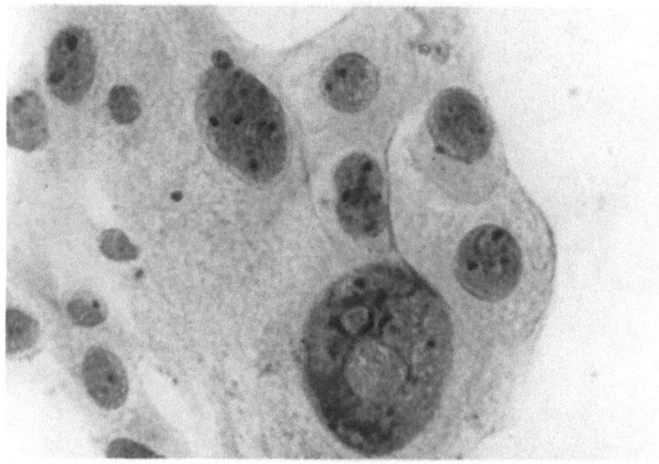

Fig. 119. Appearance of a prostatic carcinoma after irradiation 15 months previously. Signs of marked regression: giant nuclei with nuclear vacuolation; considerable rarefaction of the nuclei, with cytoplasmic lakes; small nucleoli. Oil immersion, ×540

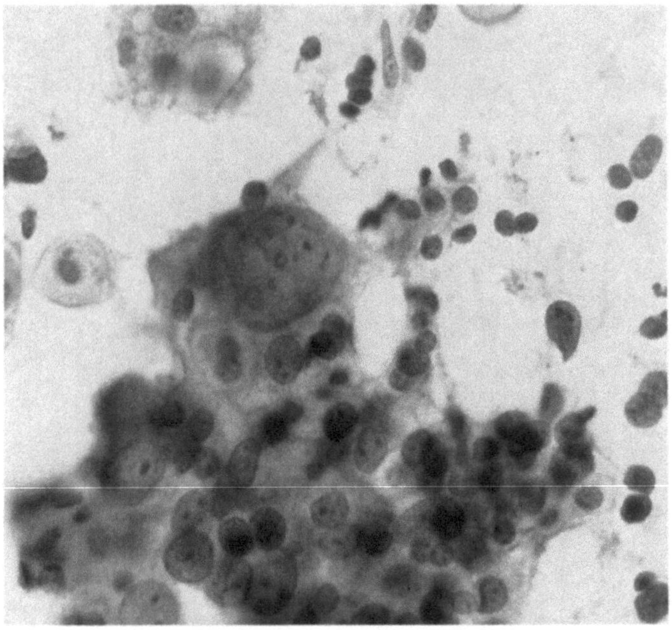

Fig. 120a. Appearance after 8 years of hormone therapy: giant nuclei at the edge of a group of cells that has undergone marked regression, with loosening of the chromatin structure and clearly defined nuclear membranes; small nucleoli. ×400

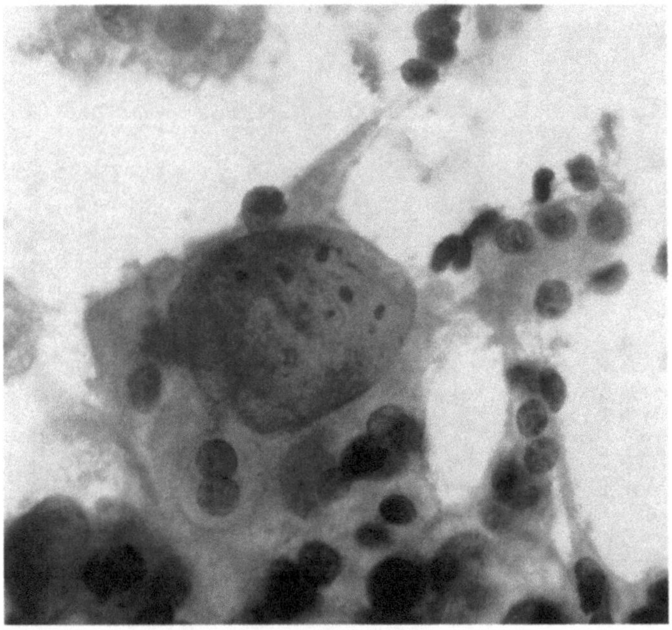

Fig. 120b. Same case at higher magnification. ×630

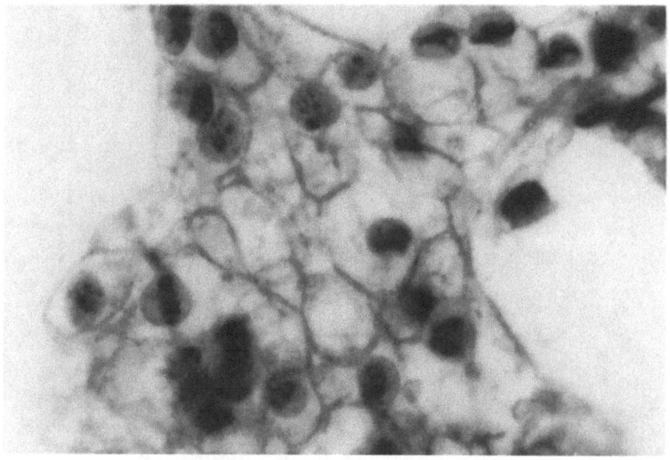

Fig. 121 a. Appearance after hormone therapy for 1 year, with coarse vacuolation of the cytoplasm and rarefaction of the nuclei; the nucleoli, however, are still prominent. Thus overall there is only slight regression. ×400

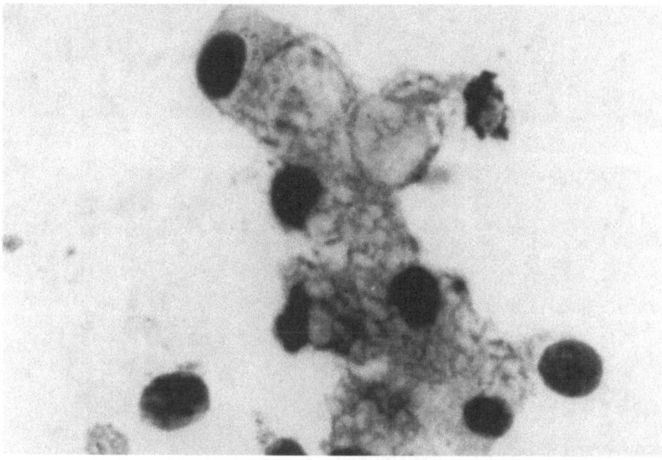

Fig. 121 b. Same case, another group of cells: fine and coarse cytoplasmic vacuoles, large nuclei. Only qualified assessment of the chromatin structure is possible. ×400

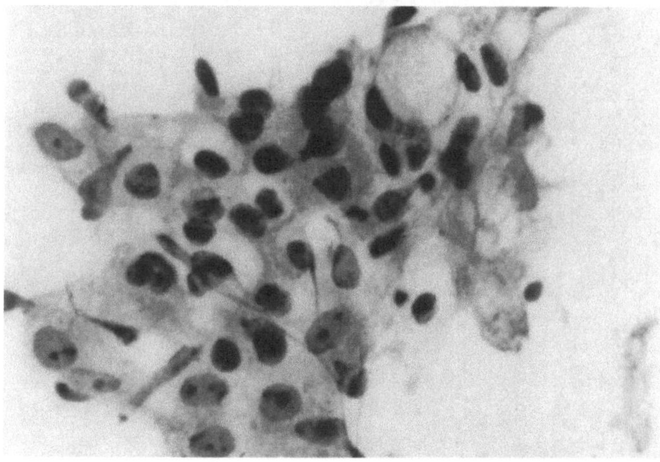

Fig. 122. Appearance after 3 months of secondary hormone therapy following unsuccessful irradiation: several coarse cytoplasmic vacuoles; the nuclei still contain numerous prominent nucleoli, and overall the regression is therefore only slight. ×400

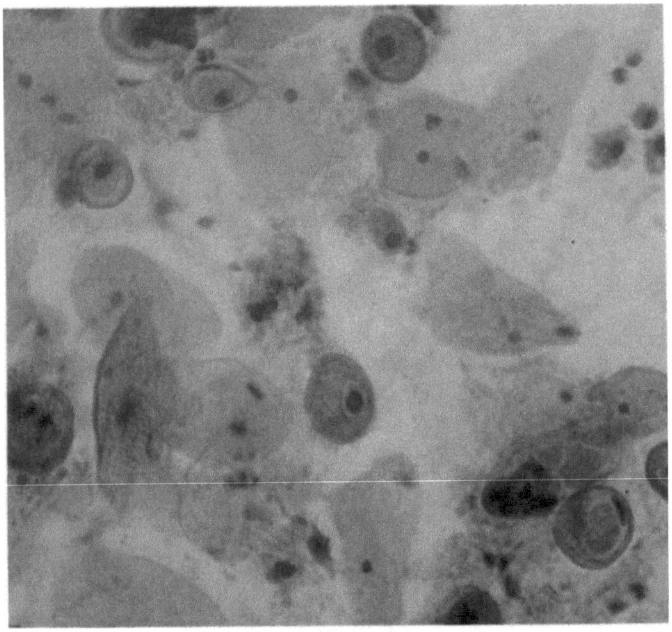

Fig. 123. Appearance after hormone therapy for 6 months: massive collection of squamous cells as a result of estrogen-induced metaplastic transformation of normal prostatic epithelial cells. Oil immersion. × 540

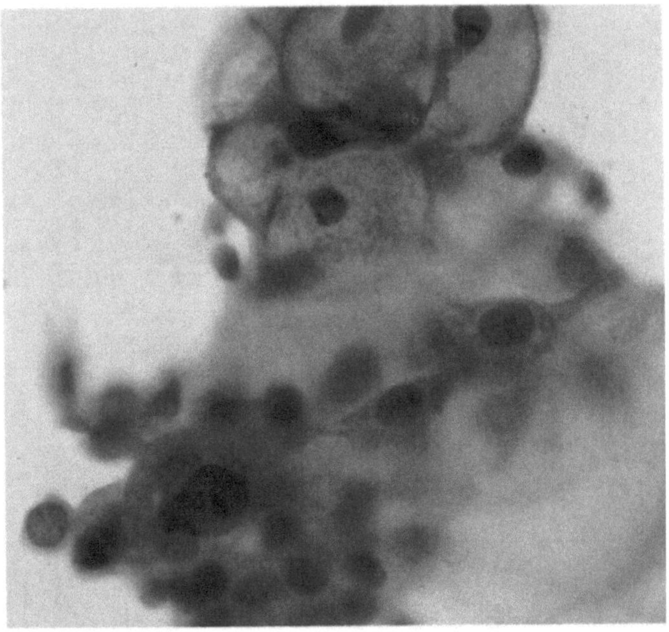

Fig. 124. Large group of metaplastically transformed prostatic epithelial cells in the upper part of the picture. × 630

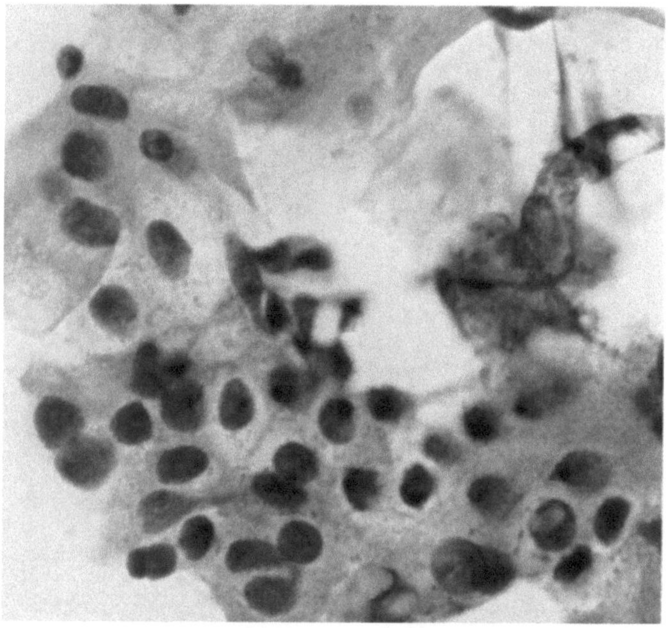

Fig. 125a. Appearance after hormone therapy for 3 months: large group of prostatic carcinoma cells in part showing regressive changes (small nucleoli!), though the chromatin structure is still dense. Isolated large squamous cells arising from metaplastically transformed benign prostatic epithelium present in the same group of cells. ×630

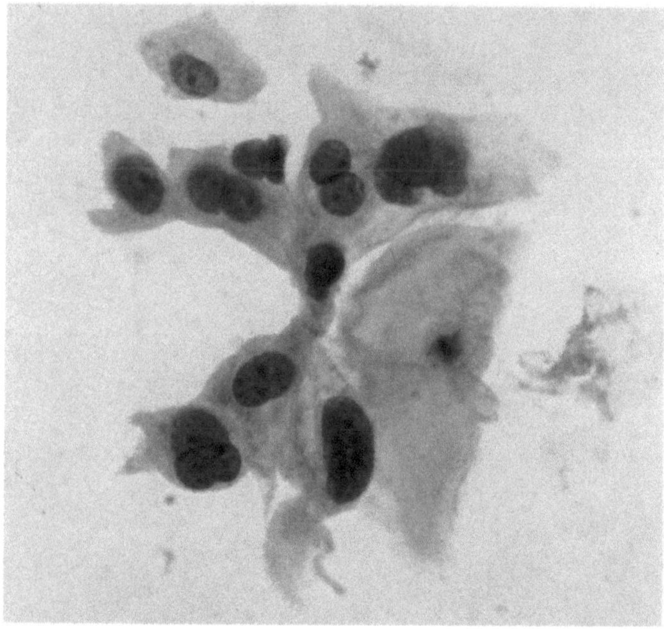

Fig. 125b. Same case, a different group of cells. Squamous cells adjoining carcinoma cells displaying only very slight regression. ×630

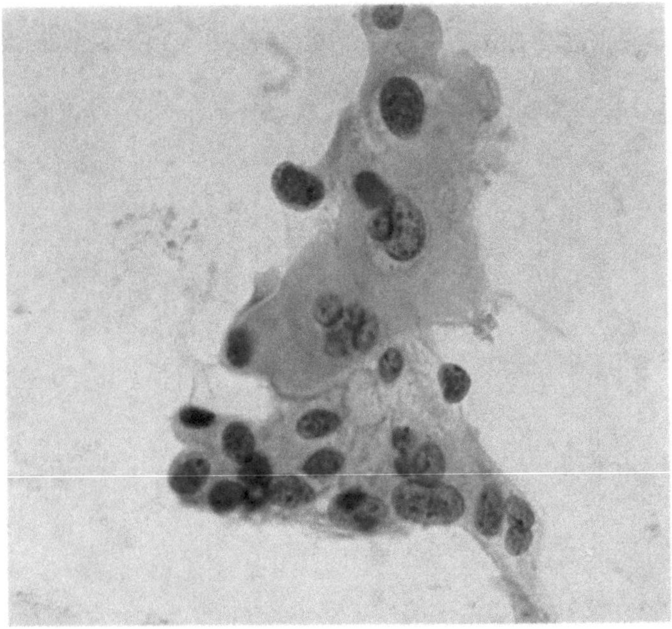

Fig. 126a. Stage of metaplastic transformation after hormone therapy for 3 months: differentiation from malignant epithelial cells is difficult; the nuclei of the metaplastic cells are markedly polymorphic and hyperchromatic; the chromatin structure is granular. The nucleoli are nevertheless only conspicuous. ×400

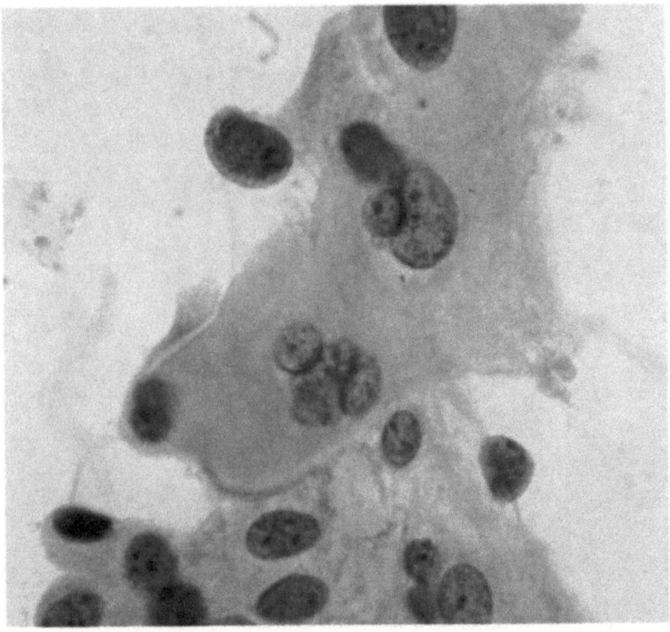

Fig. 126b. Same case at higher magnification. The lack of prominence of the nucleoli is now more evident. Reliable assessment of regression is not possible on the basis of such a group of cells. ×630

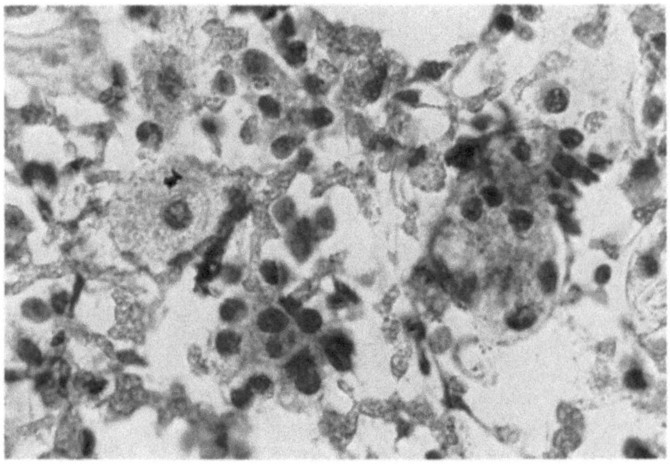

Fig. 127a. Appearance after hormone therapy for 3 weeks: variegated picture with small, in part dissociated groups of prostatic epithelial cells, isolated squamous cells and histiocytes (*upper left*). ×400

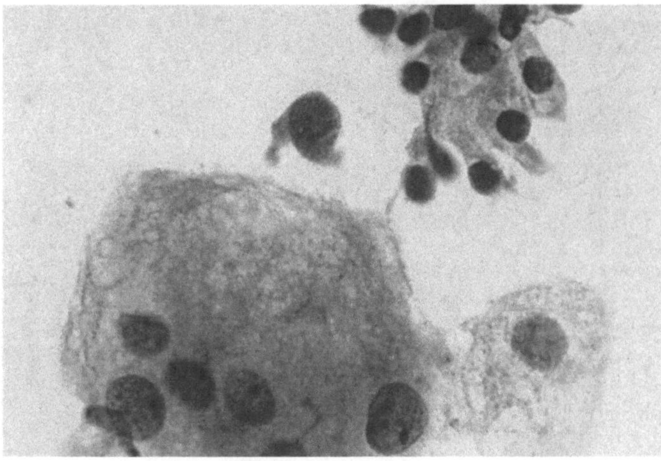

Fig. 127b. Same case: small group of prostatic epithelial cells, which cannot be classified as carcinomatous, alongside a large histiocyte with abundant cytoplasm. ×630

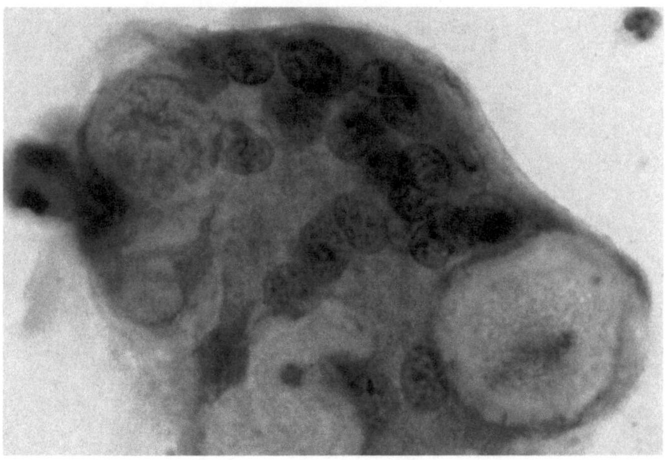

Fig. 128a. Appearance after hormone therapy for 3 years: large histiocyte with phagocytized squamous cells. ×630

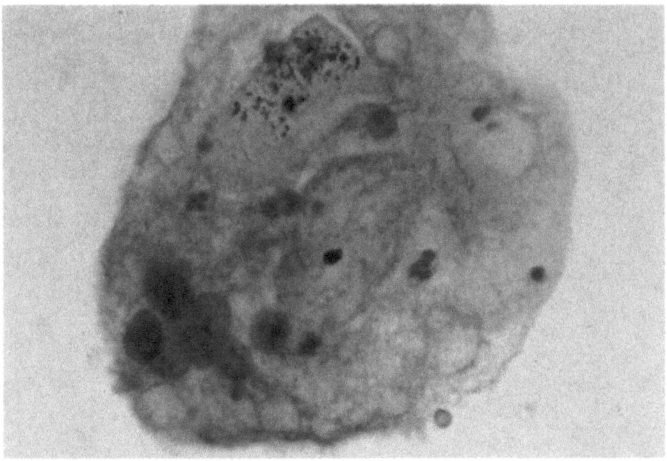

Fig. 128b. Same case: large histiocyte with foamy vacuolated cytoplasm, phagocytized squamous cells and small nuclear fragments. × 630

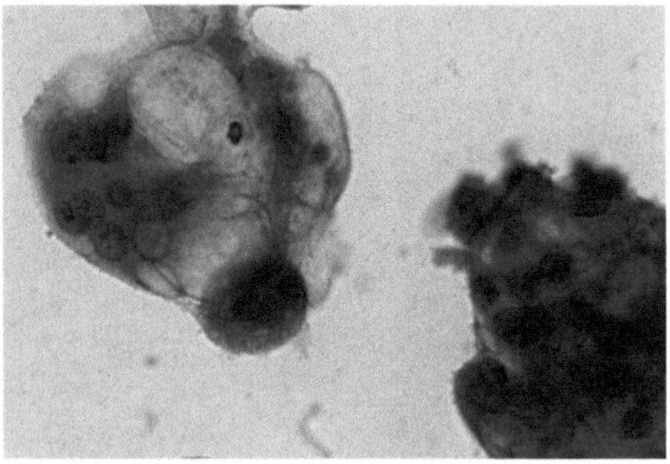

Fig. 128c. Same case: histiocyte with particularly coarse vacuolation. × 400

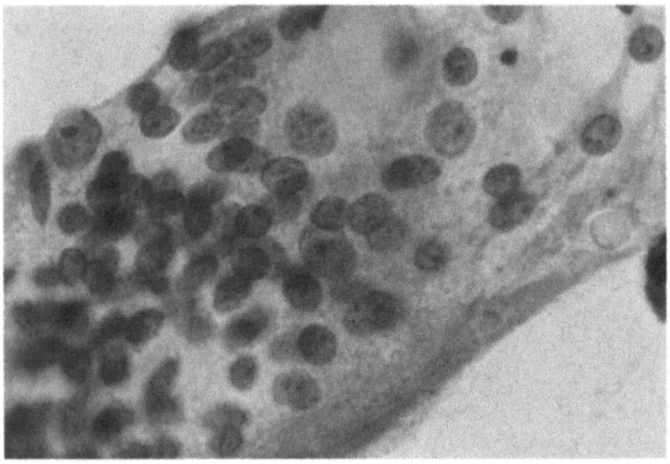

Fig. 129a. Appearance after 3 months of *Estracyt/Emcyt* therapy: histiocytic, multinucleated giant cell. Oil immersion, × 540

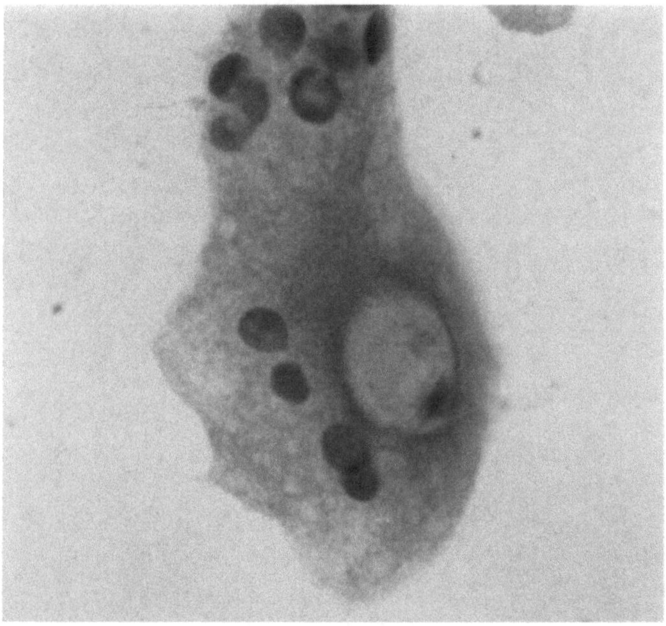

Fig. 129b. Same case: large histiocyte with relatively few nuclei and phagocytized squamous cells. Oil immersion, × 540

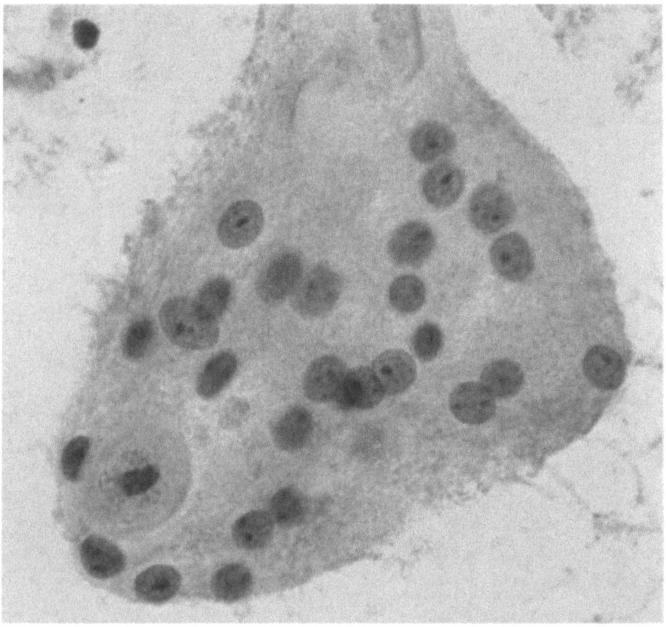

Fig. 130. Appearance after hormone therapy for 4 years: large, multinucleated histiocyte with phagocytized squamous cells. The clearly defined nuclear membrane and the loosely granular chromatin structure are, along with the phagocytosis, clear indications that this is a histiocyte despite the conspicuous to almost prominent nucleoli. Oil immersion, × 540

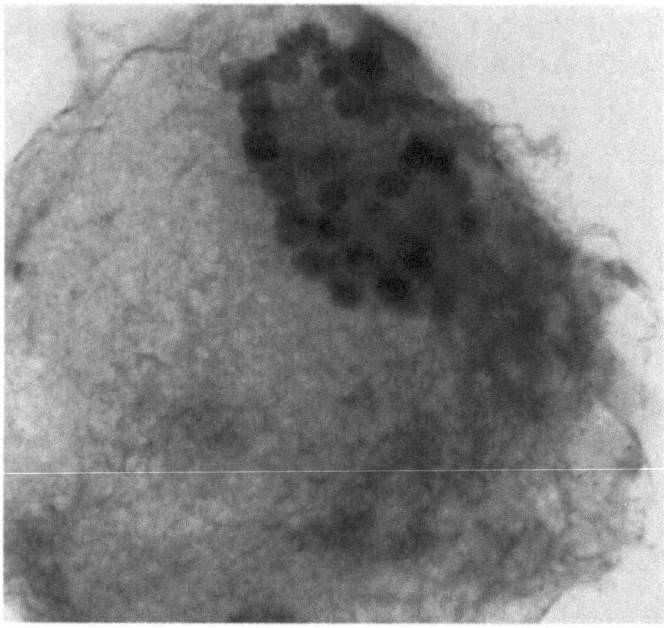

Fig. 131. Appearance after hormone therapy for 6 months: giant, multinucleated histiocyte with peripherally located nuclei and a foamy granular cytoplasmic structure. ×400

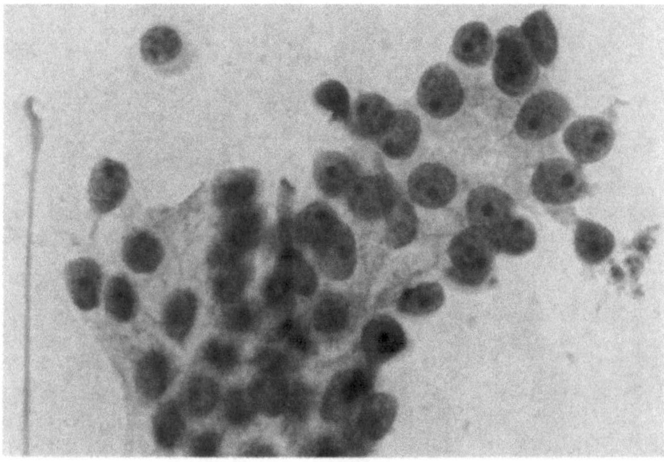

Fig. 132a. Grade II carcinoma prior to hormone therapy. ×400

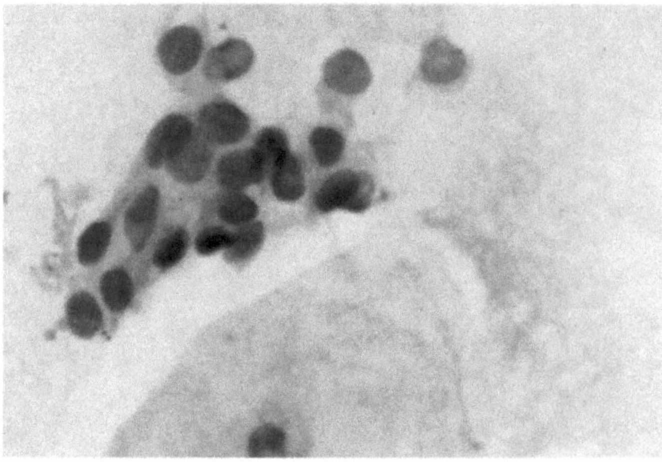

Fig. 132b. Same case after 12 months of hormone therapy: small group of prostatic epithelial cells; atypias corresponding to Pap III; classification as carcinomatous not definitely possible. Large squamous cell. Evaluated as regression grade II. ×400

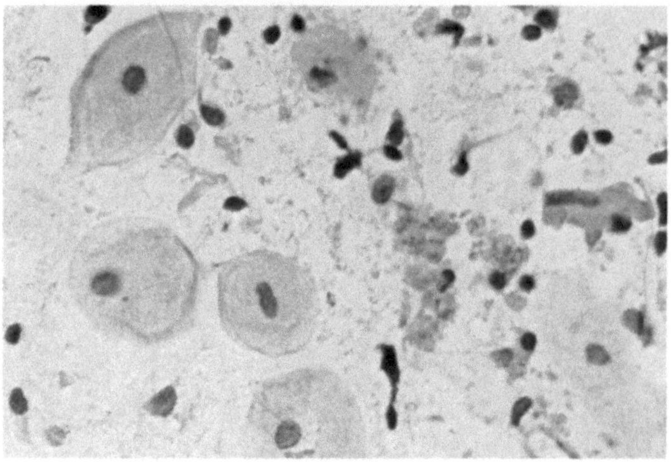

Fig. 132c. Same case: several squamous cells as an expression of therapy-induced metaplasia. ×400

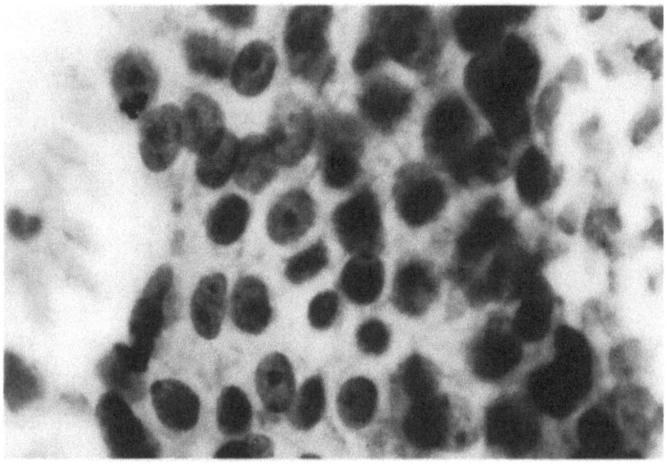

Fig. 133a. Grade II carcinoma prior to hormone therapy. ×630

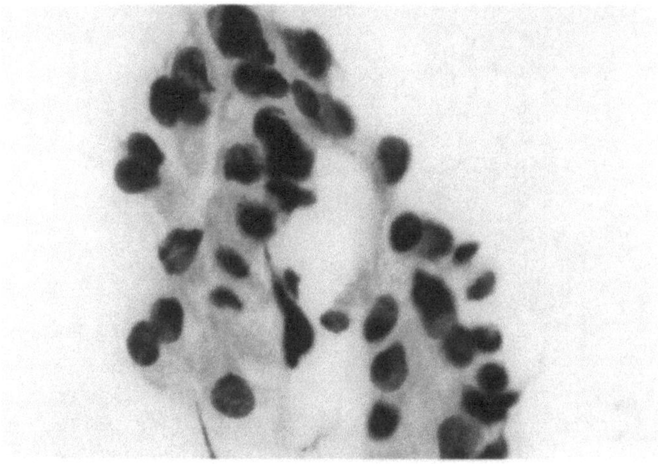

Fig. 133b. Same case after hormone therapy for 6 months: numerous nuclei are markedly smaller and most nucleoli are no longer distinguishable; there are, however, isolated nuclei with slightly conspicuous nucleoli and increased chromatin density. Overall, regression grade IV. ×400

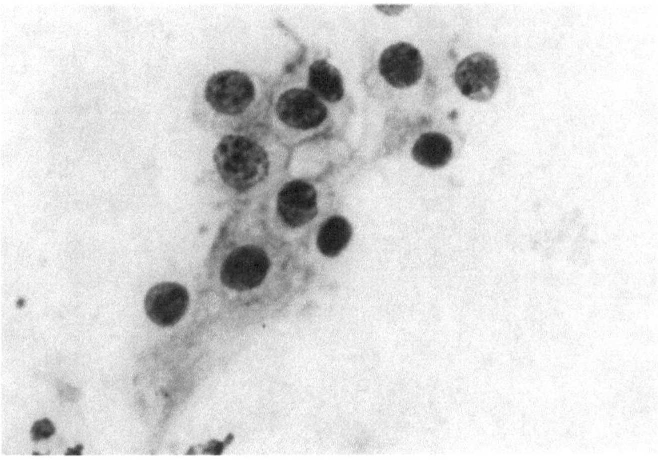

Fig. 133c. Same case: nuclei smaller than prior to therapy, and only a few still have prominent nucleoli. Cytoplasmic vacuoles. ×400

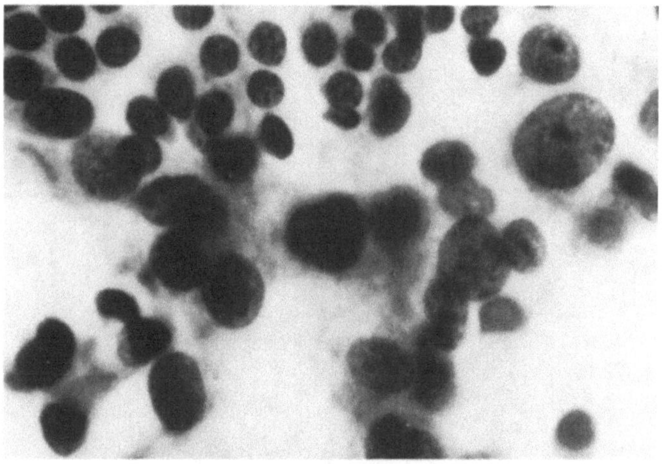

Fig. 134a. Appearance after irradiation 2 years previously: in the *upper part of the picture* unsuspicious prostatic epithelium is seen. In the *lower part of the picture* there are several nuclei with increased chromatin density but small nucleoli. *Right,* giant nuclei. Regression grade IV. ×630

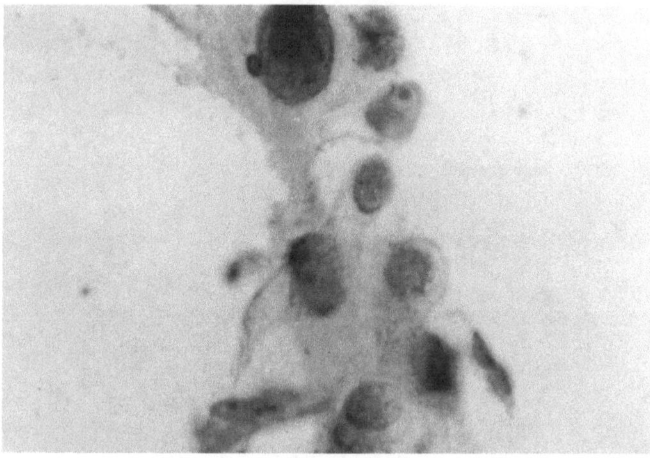

Fig. 134b. Same case: a small group of prostatic epithelial cells which cannot be classified as carcinomatous with certainty. Regression grade II. Taking into consideration Fig. 134a, however, overall classification as regression grade IV. ×630

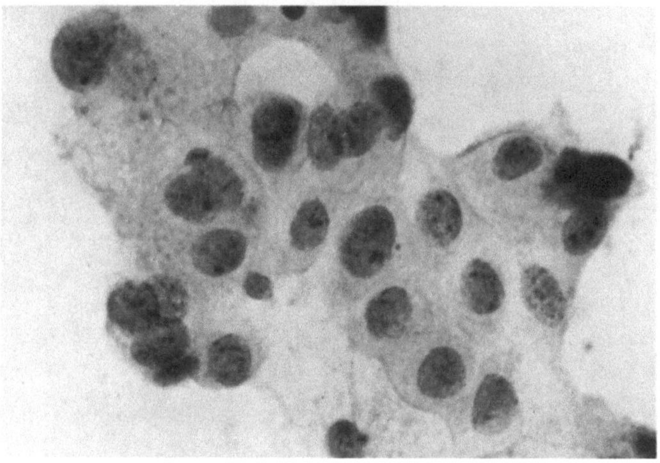

Fig. 135. Appearance after irradiation 1 year previously: prostatic epithelium with nuclei of varying size and moderately condensed chromatin; nevertheless, the nucleoli are only locally prominent, otherwise merely conspicuous. Regression grade IV. ×630

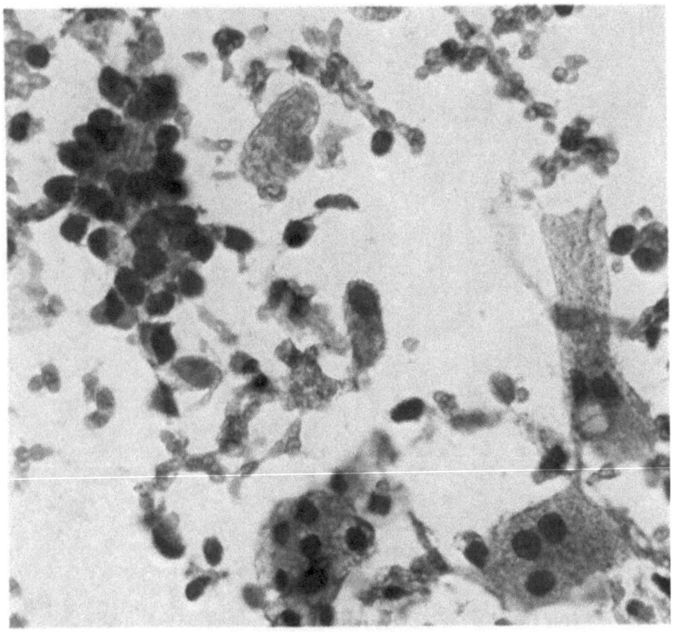

Fig. 136. Appearance after 3 weeks of hormone therapy, with small groups of cells from a residual carcinoma which even at this time shows marked regression. Regression grade IV. ×400

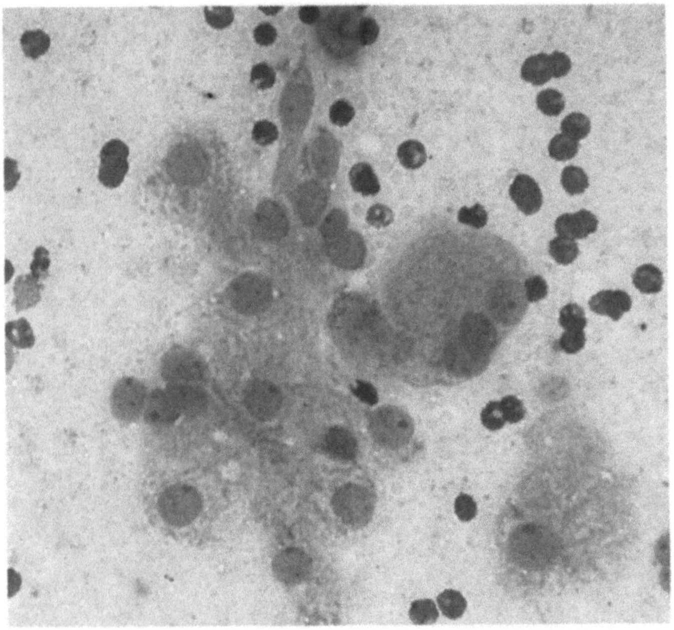

Fig. 137. Findings in regression grade IV after MGG staining. ×400

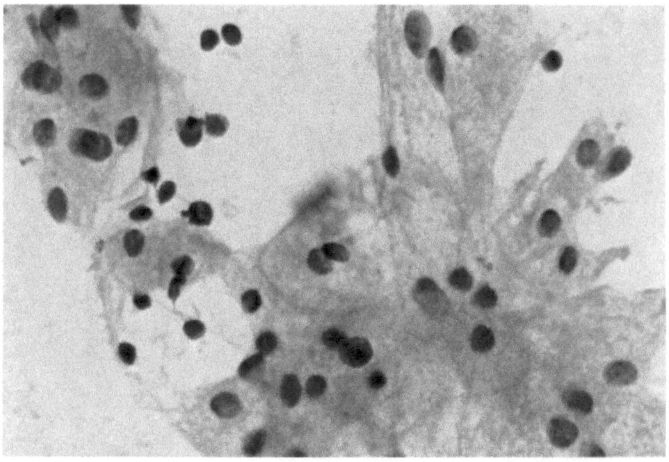

Fig. 138a. Appearance after hormone therapy for 1 year: residual carcinoma showing marked regression, with cytoplasmic lake formation. Regression grade IV. ×400

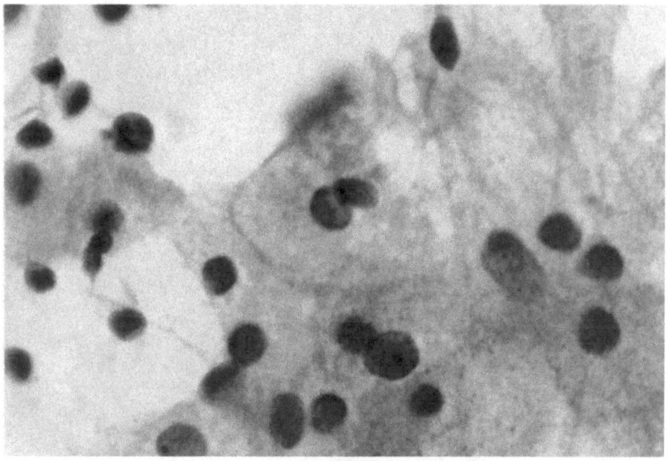

Fig. 138b. Same case at higher magnification. The still marked chromatin density, the moderate nuclear polymorphism and individual nuclei with conspicuous nucleoli are indicative of regression grade IV. ×630

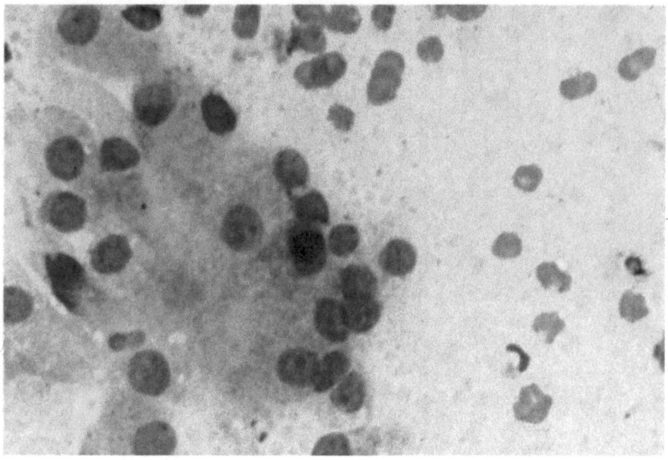

Fig. 138c. Same case after MGG staining: for technical reasons related to the staining procedure the nuclei appear larger. ×400

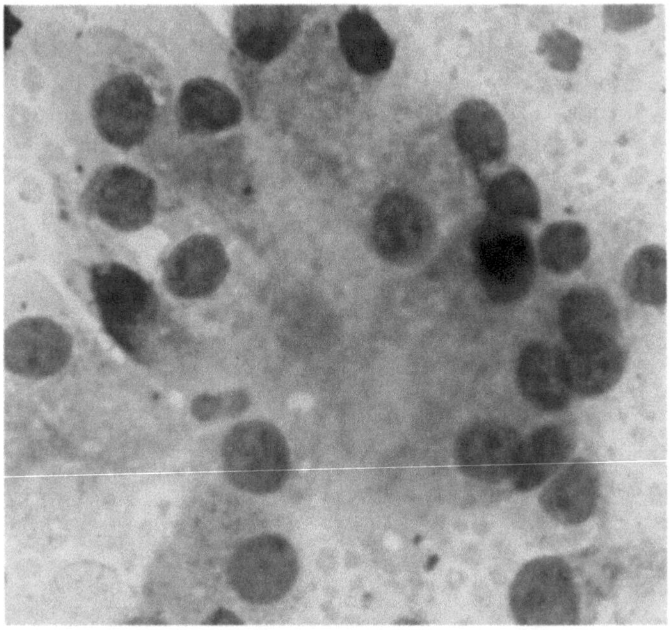

Fig. 138d. Same case at higher magnification. × 630

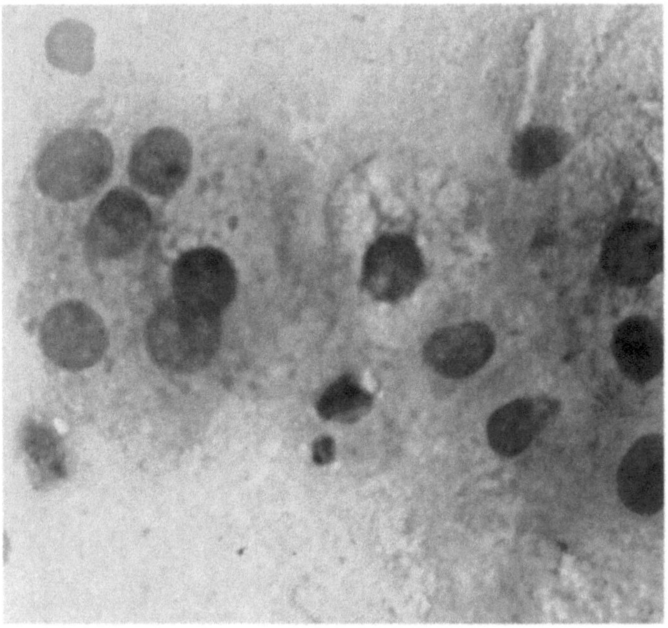

Fig. 138e. Same case. × 630

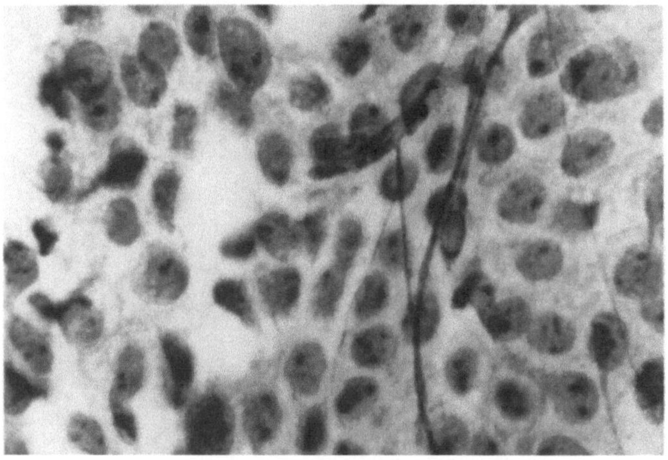

Fig. 139a. Grade III carcinoma prior to *Estracyt/Emcyt* therapy. ×400

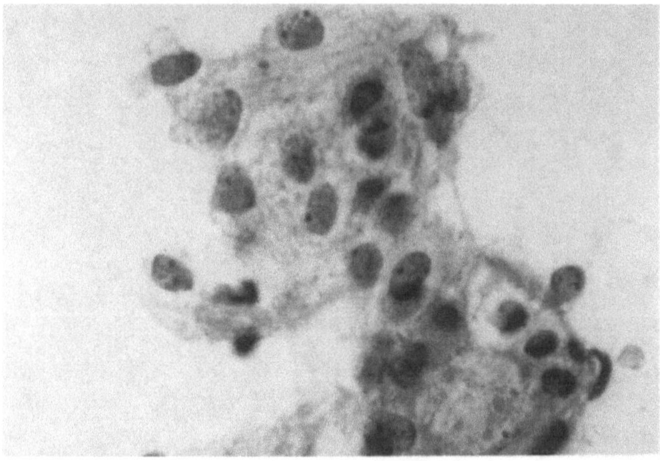

Fig. 139b. Same case after 3 months of *Estracyt/Emcyt* therapy: nuclei mostly smaller, nucleoli conspicuous or slightly prominent, nuclear rarefaction, isolated pyknotic nuclei, cytoplasmic vacuolation. Overall, regression grade VI. ×400

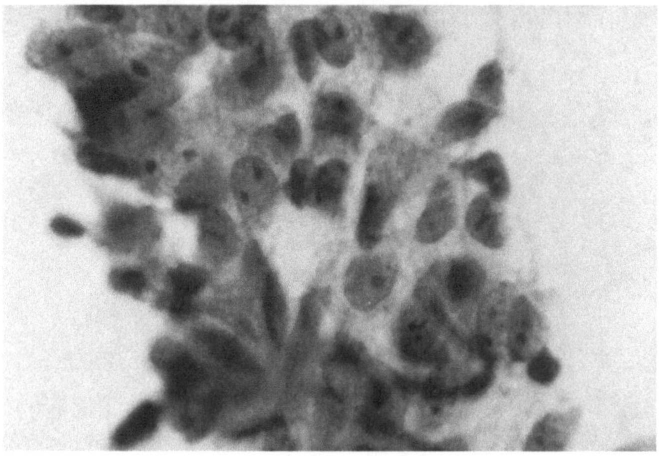

Fig. 140a. Grade III carcinoma prior to primary *Estracyt/Emcyt* therapy. ×400

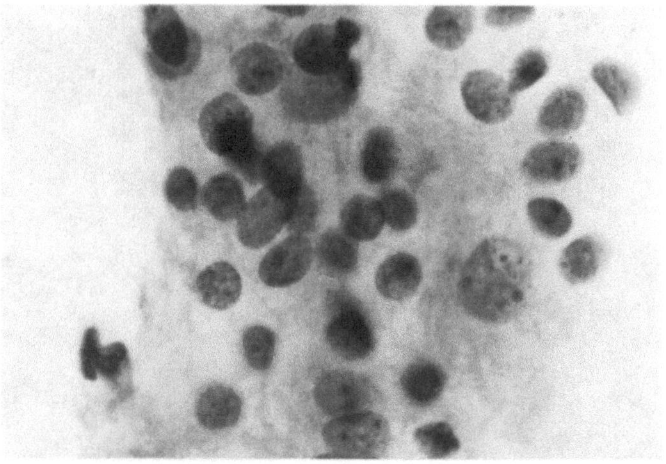

Fig. 140 b. Same case after primary *Estracyt/Emcyt* therapy for 3 months. Nuclei moderately polymorphic with a still dense chromatin structure and isolated prominent nucleoli. Most nuclei, however, are distinctly smaller and rarefied and mostly contain small nucleoli. Regression grade VI. ×400

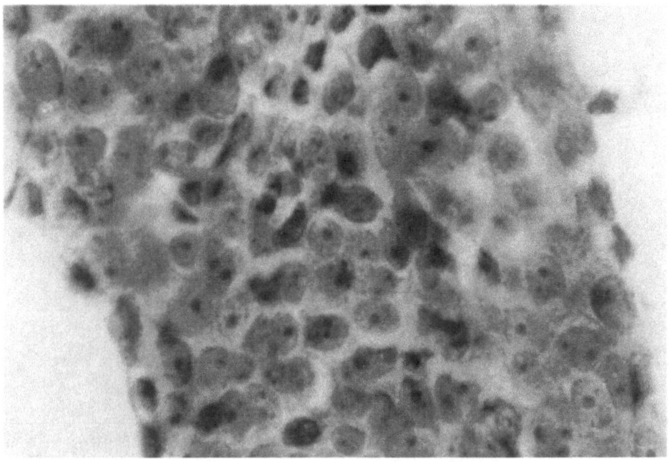

Fig. 141 a. Appearance after irradiation $1^1/_2$ years previously: grade II carcinoma without signs of regression, corresponding to regression grade X. ×400

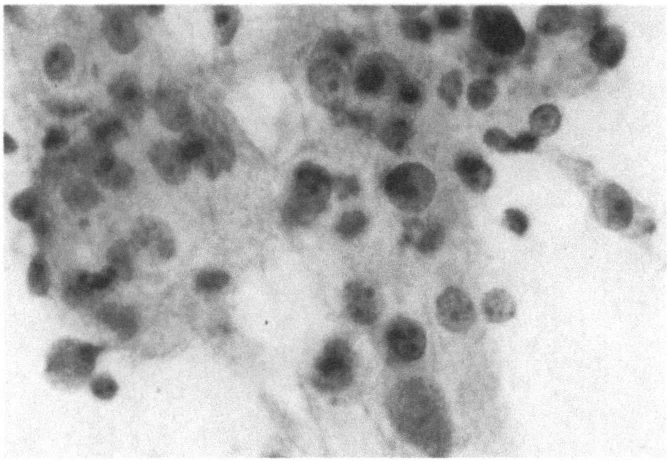

Fig. 141 b. Same case after 6 months of secondary hormone therapy: signs of marked regression are now evident in the form of predominantly smaller nuclei, nuclear rarefaction and nucleoli that are mostly only conspicuous or even scarcely recognizable. Localized cytoplasmic vacuolation. Overall, regression grade VI. ×400

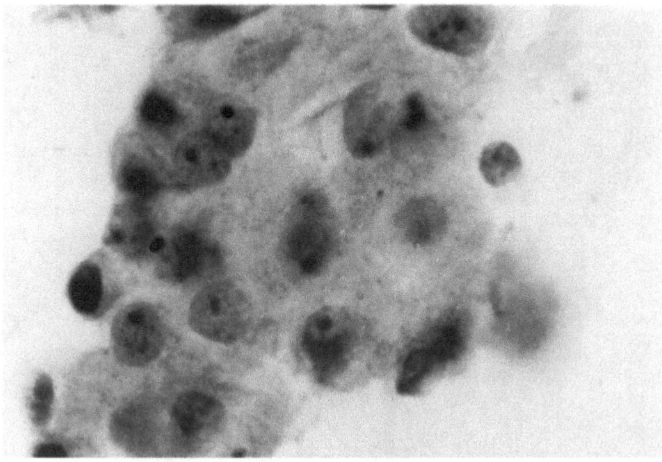

Fig. 142a. Appearance after irradiation 2 years previously: group of prostatic carcinoma cells displaying signs of only slight regression corresponding to regression grade VIII. ×400

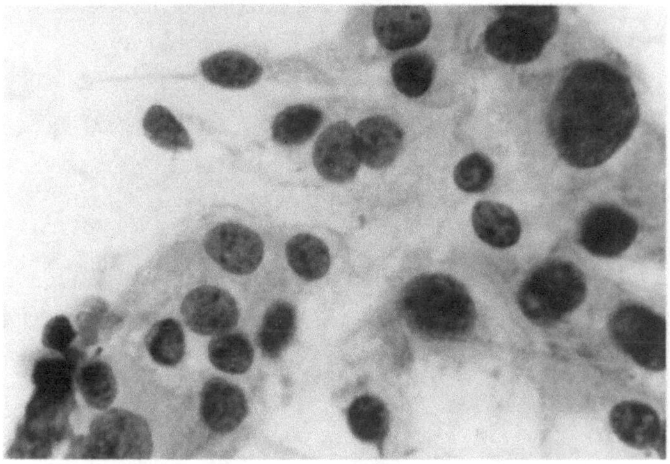

Fig. 142b. Same case after 6 months of secondary hormone therapy: most nuclei are markedly smaller and rarefied. The nucleoli are now only conspicuous or scarcely distinguishable. A giant nucleus is present at the upper right. There is loosening and shrinkage of the cytoplasm. Overall, regression grade VI. ×400

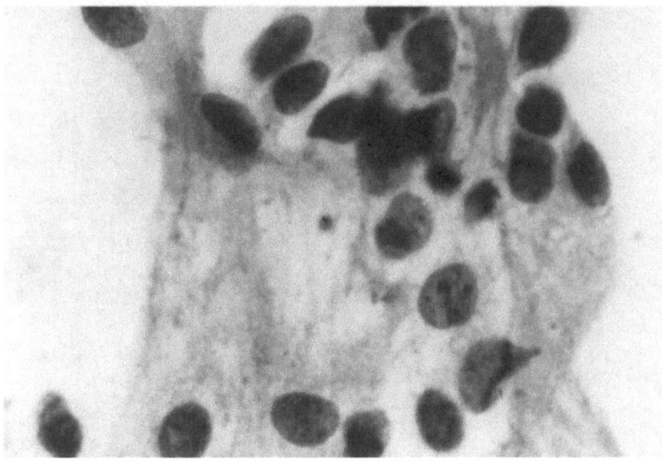

Fig. 142c. Same case, another group of cells. Signs of regression as described in Fig. 142b. ×400

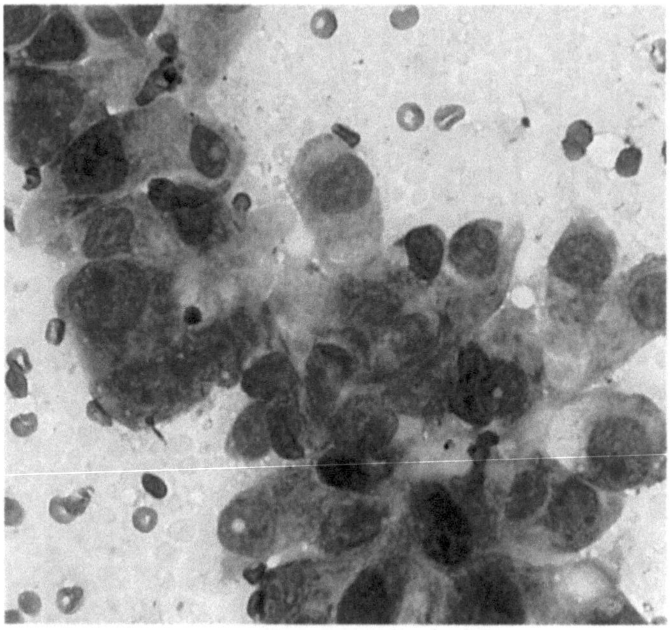

Fig. 143. Appearance after 6 months of estrogen therapy, in an MGG-stained smear: there is still pronounced nuclear polymorphism and disturbed nuclear arrangement; isolated prominent nucleoli are seen, though most nuclei contain only conspicuous or scarcely recognizable nucleoli. Cytoplasmic vacuolation. Classification as regression grade VI. ×400

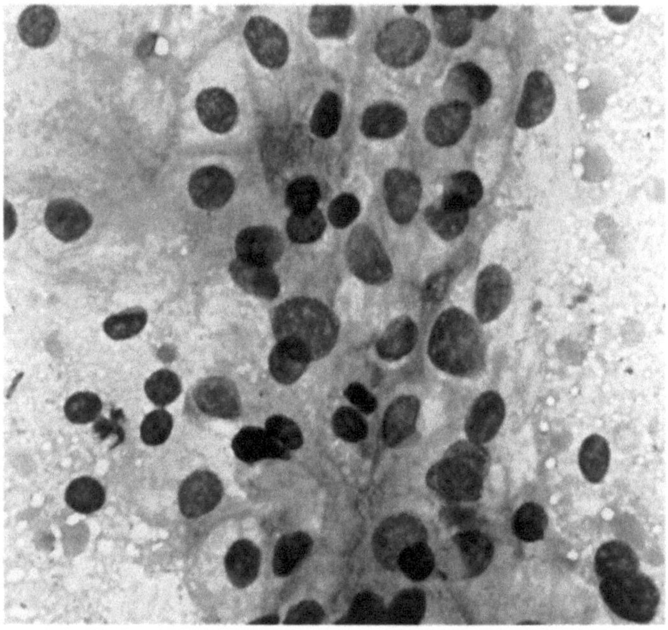

Fig. 144. Appearance after estrogen therapy and bilateral orchiectomy 1 year previously; MGG-stained smear: slight nuclear polymorphism is still present, and there is moderate disturbance of the nuclear arrangement; nevertheless, nucleoli can scarcely be detected. Cytoplasmic vacuolation. ×400

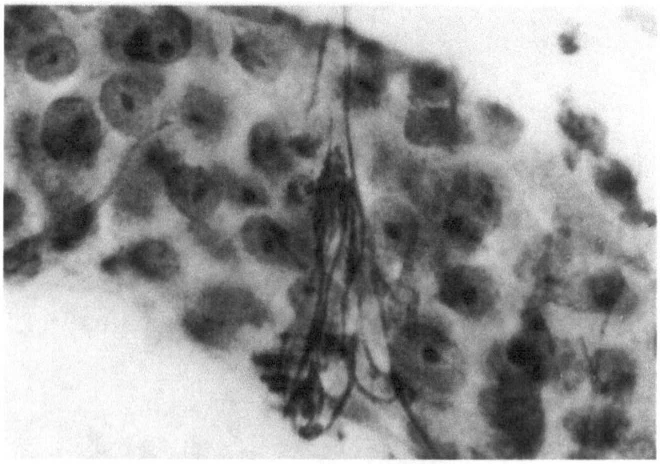

Fig. 145a. Grade III carcinoma following irradiation 6 months previously and clinically progressing metastasis. No signs of regression, corresponding to regression grade X. Oil immersion, ×540

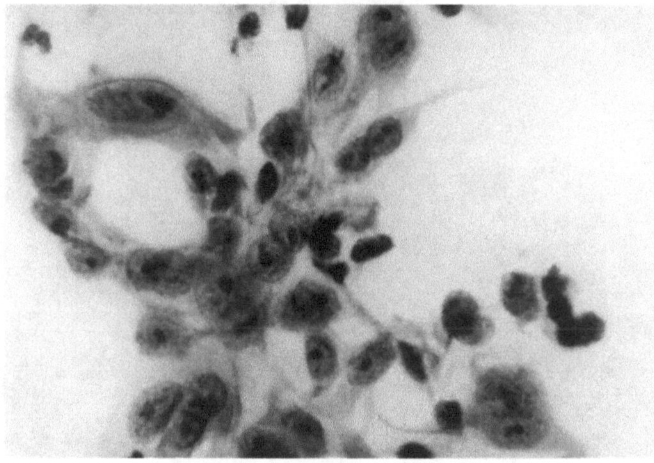

Fig. 145b. Same case after 3 months of secondary *Estracyt/Emcyt* therapy: apart from occasional reductions in nuclear size and a few pyknotic nuclei, there are no signs of regression. The nucleoli are still very prominent and polymorphic; hence regression grade VIII. Oil immersion, ×540

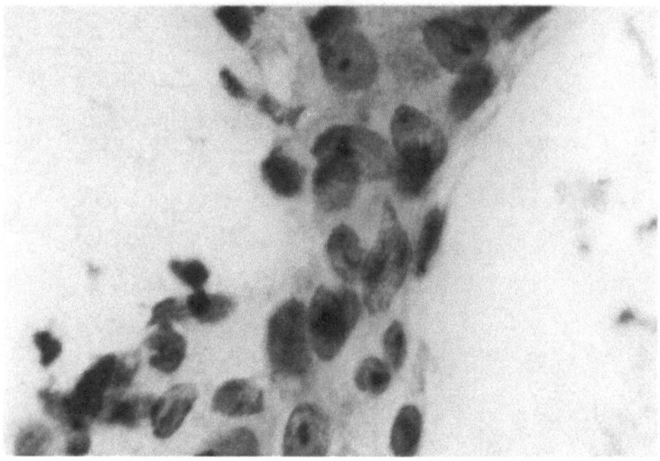

Fig. 146a. Grade III carcinoma after 6 months of primary estrogen therapy, displaying no signs of regression and corresponding to regression grade X. ×400

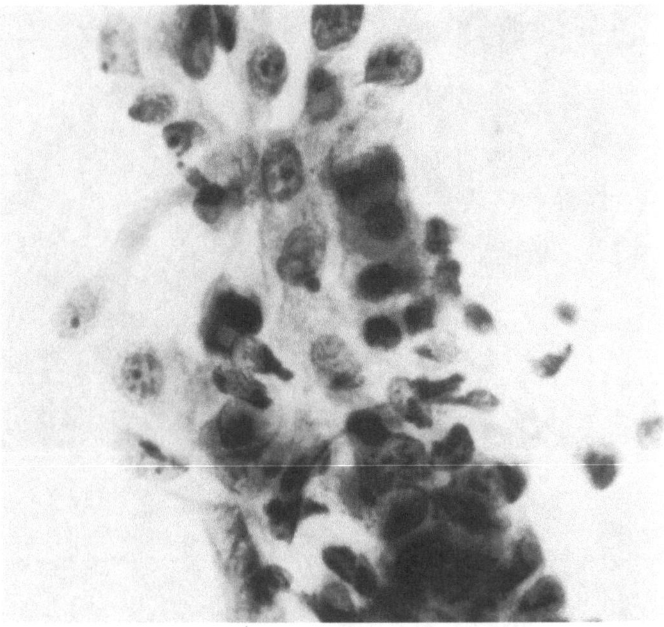

Fig. 146b. Same case after 3 months of secondary *Estracyt/Emcyt* therapy: along with signs of regression in the form of some reductions in nuclear size and pyknotic nuclei as well as low-grade cytoplasmic vacuolation, there are predominantly pronounced nuclear polymorphism, disturbance of the nuclear arrangement and prominent nucleoli. Regression grade VIII. ×400

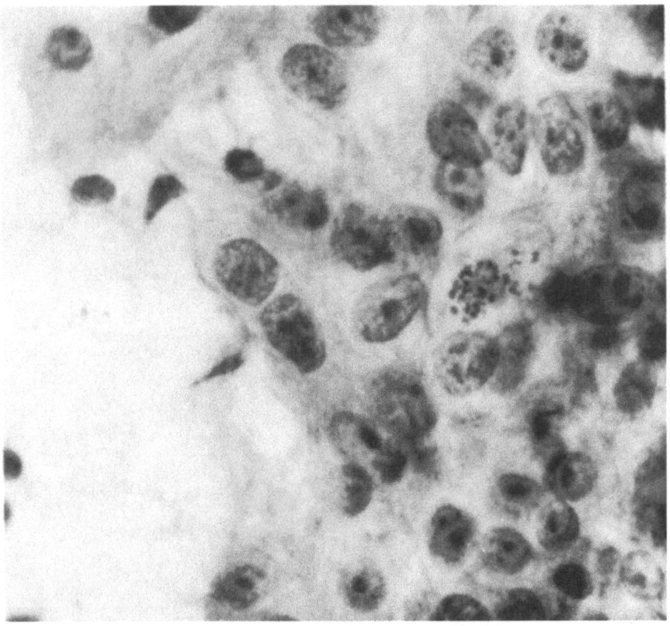

Fig. 146c. Same case after 6 months of secondary *Estracyt/Emcyt* therapy: the signs of regression are still only slight, with moderate nuclear rarefaction, diminution of isolated nuclei and cytoplasmic shrinkage. Marked nuclear polymorphism and very prominent, polymorphic nucleoli, and mitosis. Overall, regression grade VIII. ×400

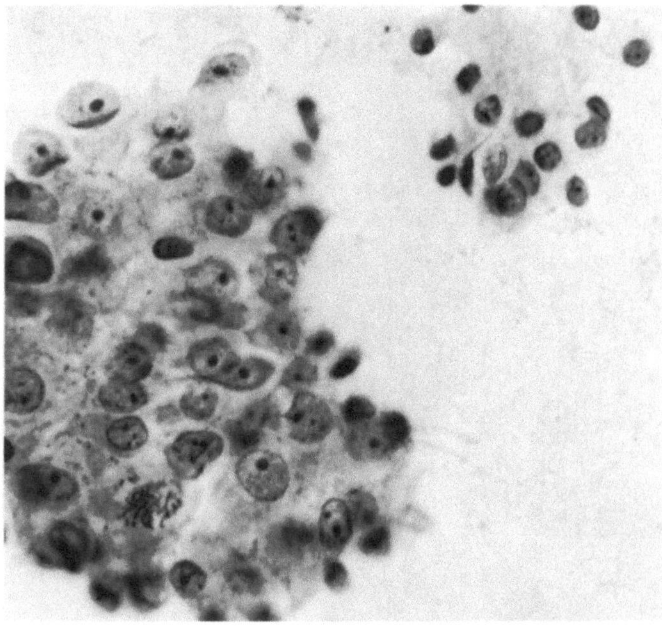

Fig. 147. Appearance after 1 year of estrogen therapy, following bilateral orchiectomy: large group of cells showing only slight signs of regression, alongside a smaller group which cannot definitely be classified as carcinomatous. It is the group displaying only slight regressive changes which is decisive for the diagnosis! ×400

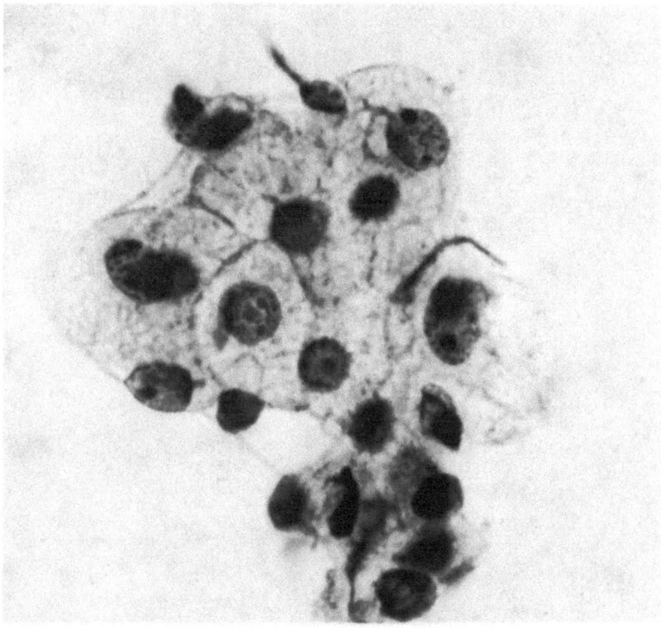

Fig. 148. Appearance after 1 year of estrogen therapy: despite considerable cytoplasmic vacuolation the nuclei still have a very dense chromatin structure and prominent nucleoli. Thus overall the regression grade is R VIII. Oil immersion, ×540

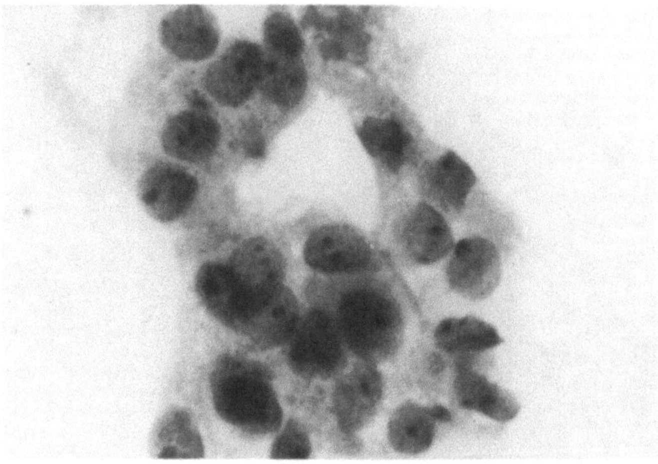

Fig. 149. Appearance after 3 weeks of estrogen therapy: only slight disaggregation of the cytoplasm and isolated nuclei with small nucleoli; most nuclei, however, contain clearly prominent nucleoli. Regression grade VIII. ×400

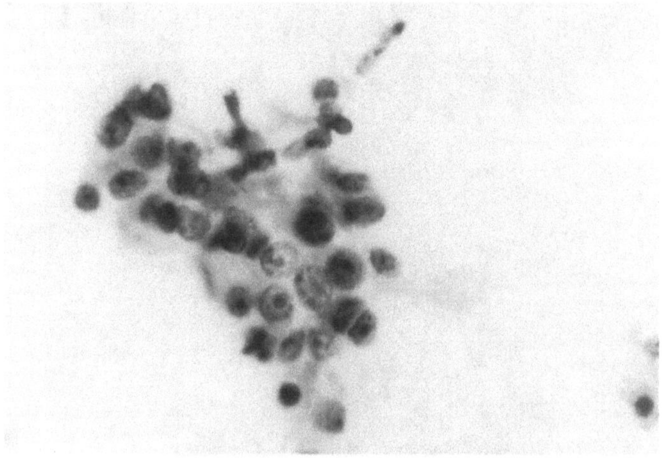

Fig. 150. Appearance after 3 years of estrogen therapy, following bilateral orchiectomy: the majority of the nuclei still display markedly prominent nucleoli, and there is nuclear polymorphism and disturbance of the nuclear arrangement. Regression grade VIII. ×400

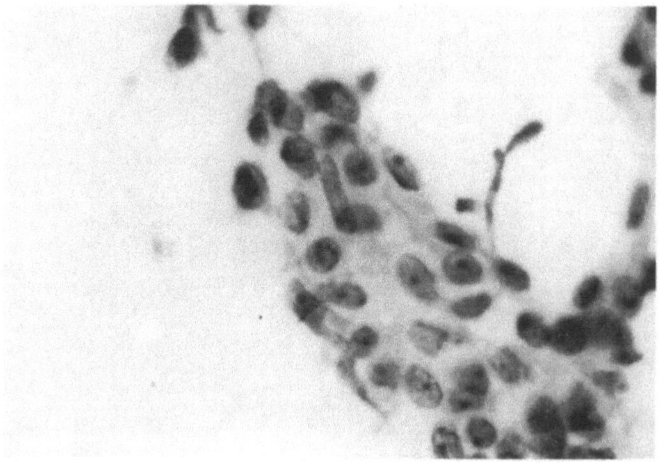

Fig. 151. Appearance after 3 months of secondary *Estracyt/Emcyt* therapy: group of carcinoma cells in which the signs of regression are predominantly slight, corresponding to regression grade VIII. ×400

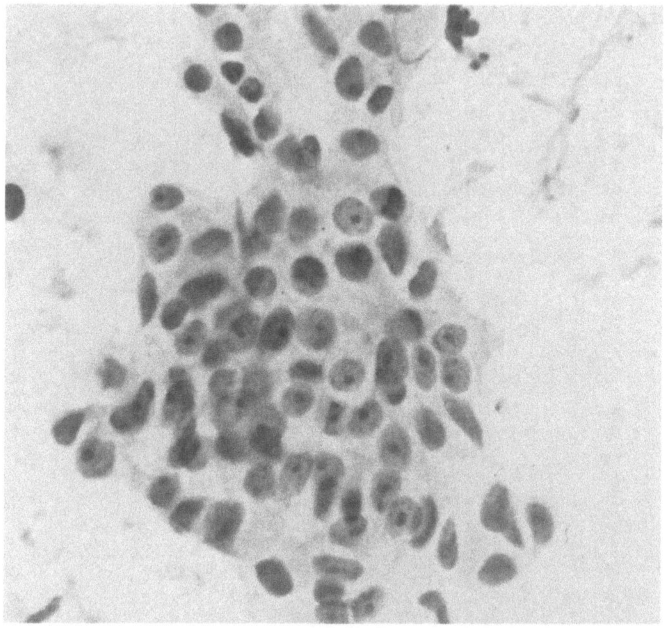

Fig. 152. Appearance after irradiation 2 years previously: group of carcinoma cells with signs of only slight regression, corresponding to regression grade VIII. ×400

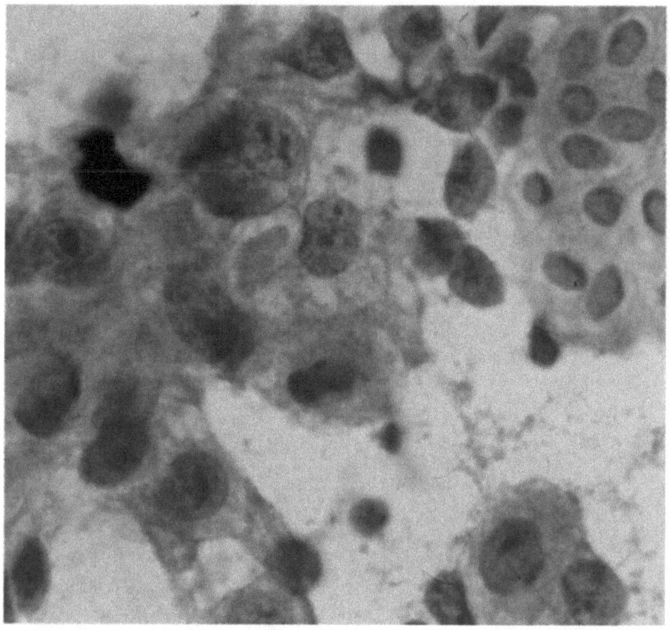

Fig. 153. Appearance after 3 years of estrogen therapy: at the *upper right of the picture* is a sheet of normal prostatic epithelium, while on the *left* are carcinoma cells displaying little regression. Overall, regression grade VIII. ×630

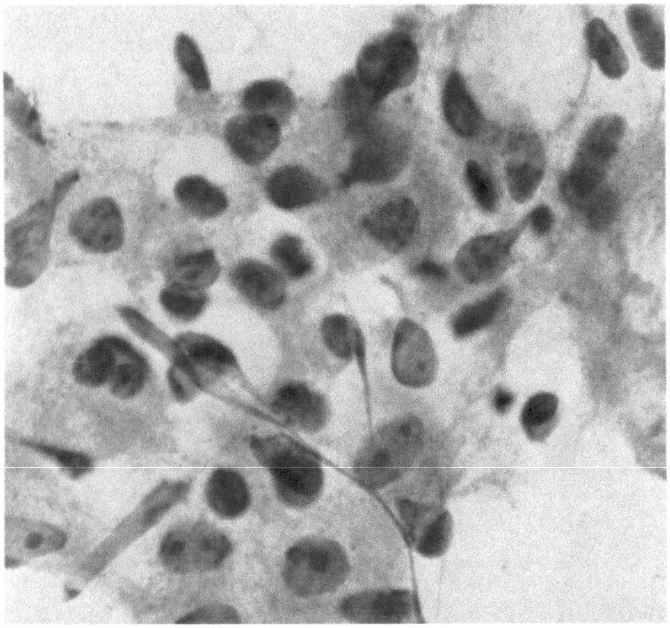

Fig. 154. Appearance after unsuccessful primary irradiation and 1 year of secondary estrogen therapy in conjunction with bilateral orchiectomy: although a few nuclei show only small nucleoli and cytoplasmic vacuolation, most still have markedly prominent nucleoli. The overall regression grade is therefore R VIII.

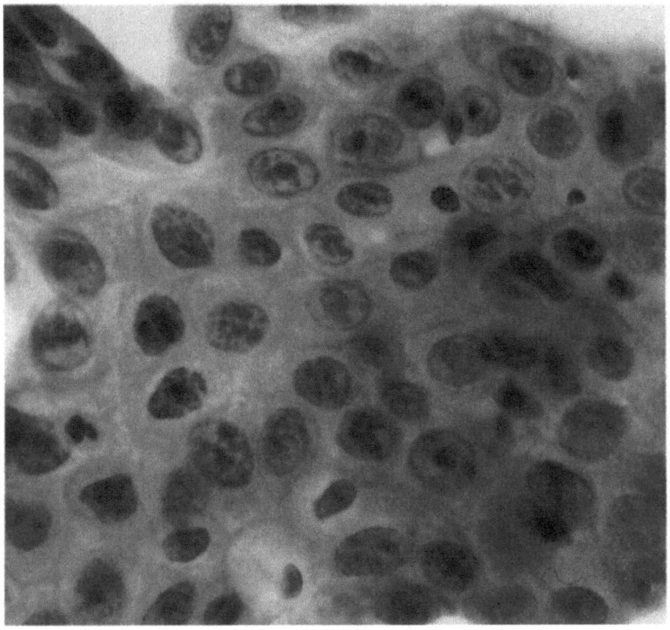

Fig. 155a. Appearance after 3 months of cytostatic treatment with Cyclophosphamide as tertiary therapy: only isolated nuclei are relatively small and have merely conspicuous nucleoli; most nuclei are large and contain very prominent and polymorphic nucleoli. Regression grade VIII. ×400

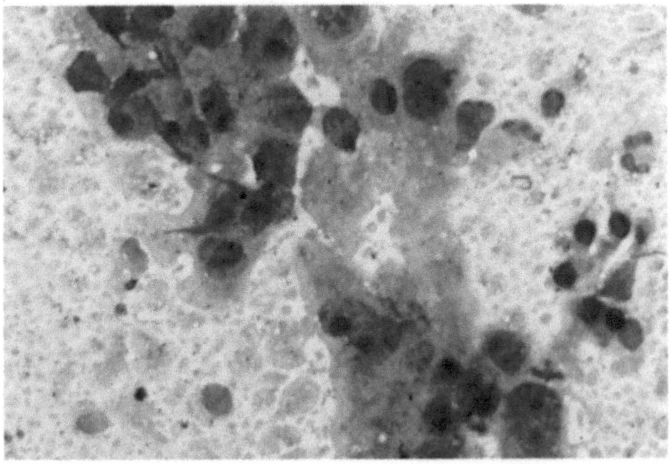

Fig. 155b. Same case, MGG stained: the nucleoli, most of which are prominent, are clearly recognizable. ×400

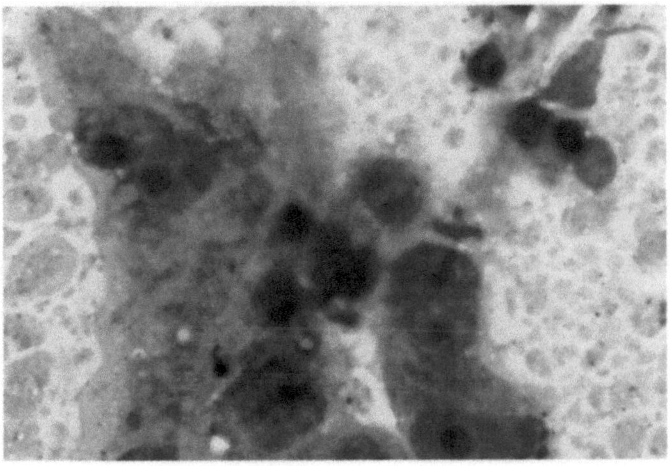

Fig. 155c. Same case at higher magnification. ×630

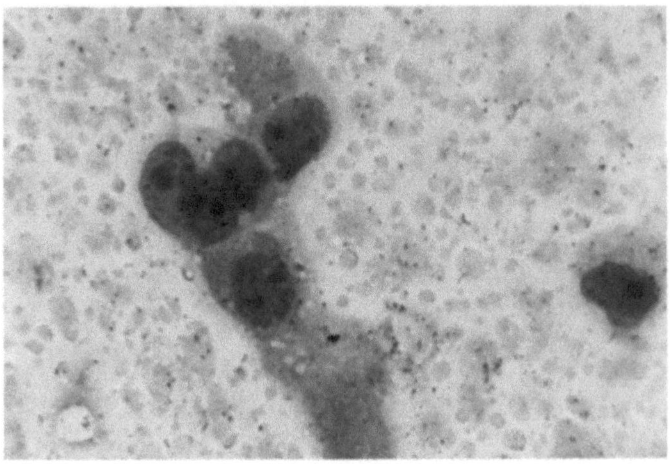

Fig. 155d. Same case: a small cluster of nuclei without signs of regression. ×400

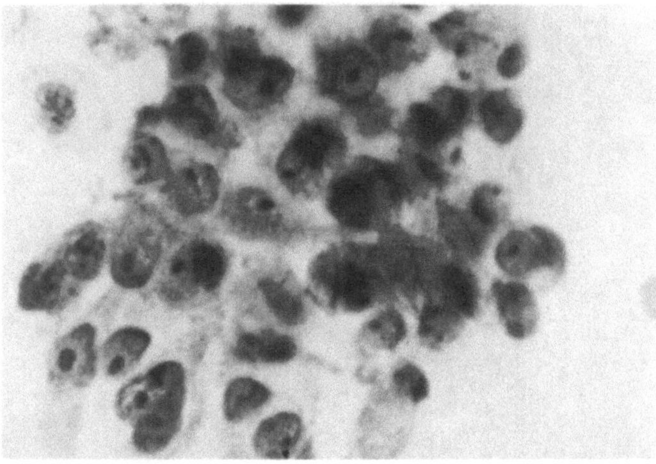

Fig. 156a. Appearance after estrogen therapy for 2 years: no signs of regression, corresponding to regression grade X. ×630

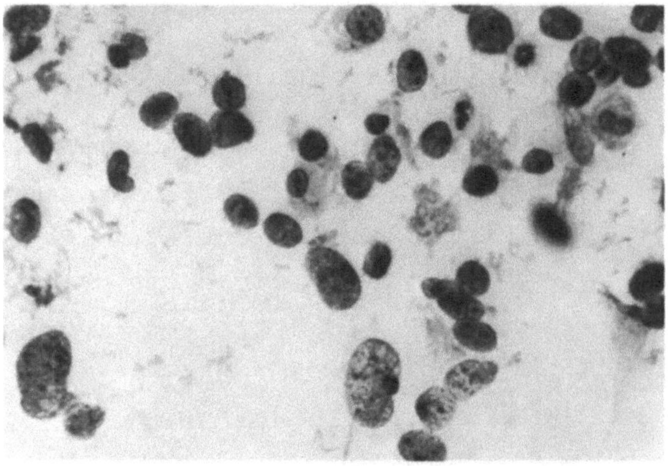

Fig. 156b. Same case, with extremely polymorphic, completely dissociated 'naked' nuclei. ×400

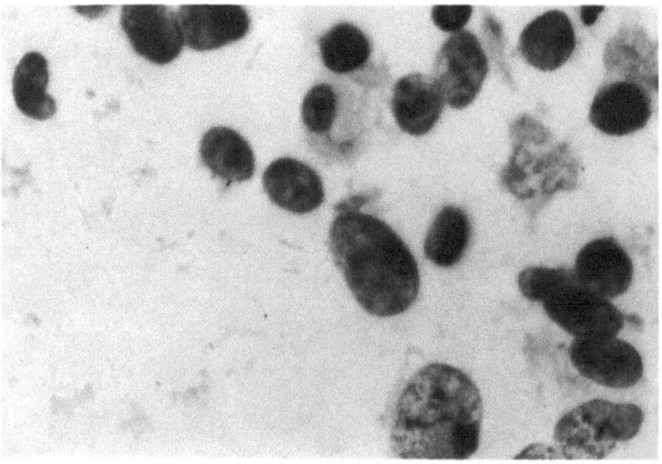

Fig. 156c. Same case at higher magnification. ×630

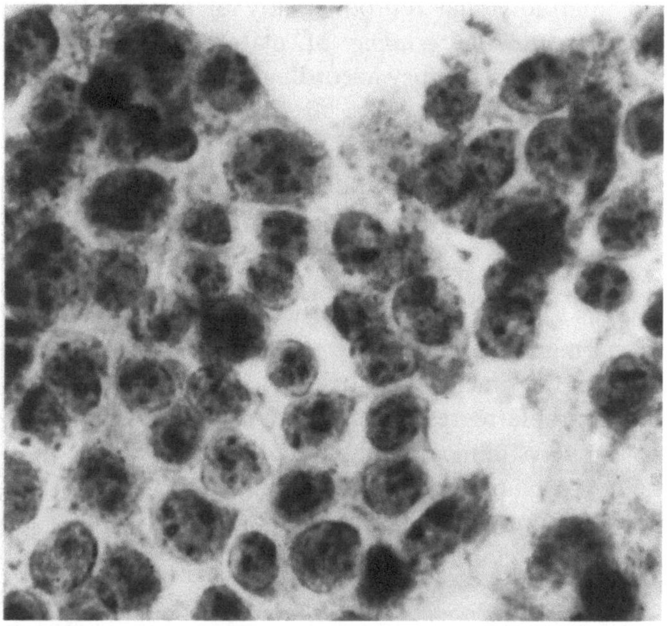

Fig. 157a. Appearance after estrogen therapy for 3 months (primary grade III carcinoma): no signs of regression. Pronounced nucleolar polymorphism. High grade of malignancy! Overall, regression grade X. ×400

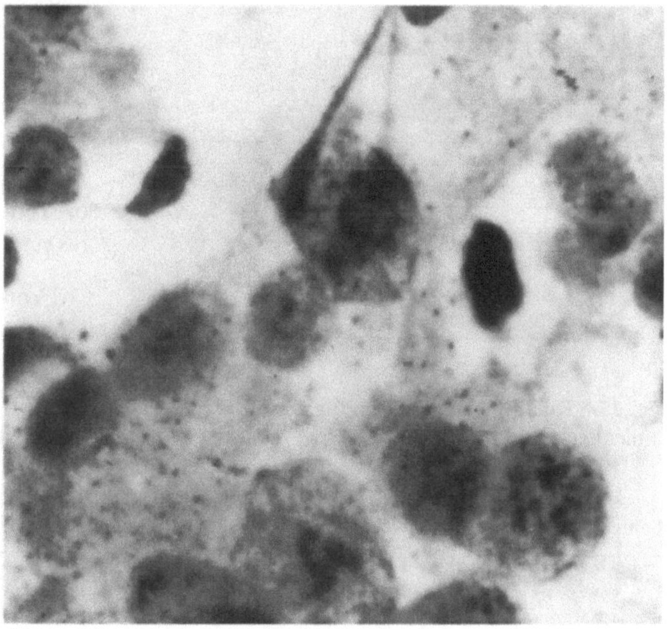

Fig. 157b. Same case after a total of 6 months contrasexual therapy, and at higher magnification: increasing high-grade nuclear anaplasia in the presence of complete hormone resistance and regression grade X. ×630

As yet no data are available on the intra- and interindividual reproducibility of *histological regression grading* of material obtained by punch biopsy according to Dhom.

9.4 Clinical Significance of Cytological Regression Grading

There is a close correlation between the cytological regression grade and the clinical condition at the time of the biopsy: patients with the favorable regression grades 0–VI are almost always clinically stable at the time of the biopsy, i.e. they exhibit no signs of tumoral progression, whereas if poor (R VIII) or no (R X) regression is observed during therapy, the rate of clinical stability is also significantly lower **(Fig. 158).**

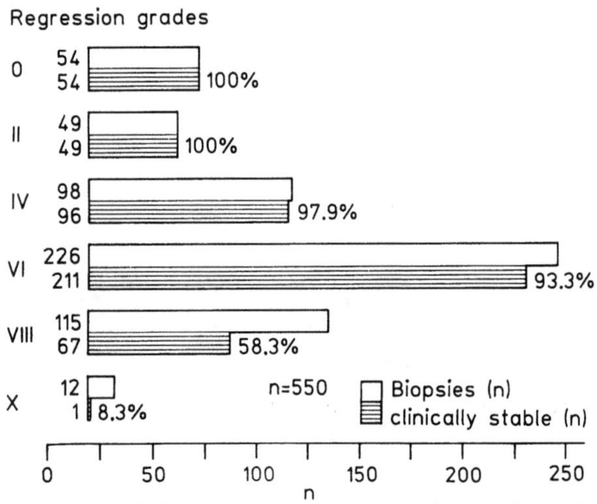

Fig. 158. Relationship between regression grades and clinical condition

Equally convincing is the correlation between the clinically demonstrated metastatic status and the simultaneously assessed regression grade (LEISTENSCHNEIDER and NAGEL 1983).

As can be seen from **Table 21,** 15 out of 16 patients (94%) with clinical evidence of metastatic regression during the treatment had cytologically favorable regression grades (R IV or VI), which substantiates the effect of the treatment on the primary tumor.

By contrast, 56 out of 66 patients (85%) with proven progression of metastasis had an unfavorable regression grade (R VIII or X). In a further eight of these patients (12%), both marked and poor regression was evident in groups of carcinoma cells; these cases were classified cytologically as regression grades VI–VIII. Simultaneously, however, increasing metastasis was observed in these eight patients, and the clinical therapeutic effect was therefore again judged to be 'poor'. Accordingly, the cytological assessment concurred with the clinical course in 64 of the 66 patients (97%).

From these investigations – and paying special regard to the latter group (R VI–VIII) – it is evident that the part of the tumor having the worst regression grade determines the clinical prognosis.

Table 21. Correlation between metastatic status and cytological regression grading

	n	Cytological regression grade				
		IV	VI	VIII	X	VI–VIII
Objectively stable	4	–	3	–	–	1
Objective regression	16	4 (25%)	11 (69%)	–	–	1 (6%)
Objective progression	66	–	2 (3%)	52 (79%)	4 (6%)	8 (12%)
	86					

9.5 Validity of Cytological Regression Grading

The absence of clinical symptoms in the locally advanced but asymptomatic stage T_3 lends particular prognostic significance to cytological regression grading, regardless of the form of treatment administered.

At this stage there is a significant correlation between the cytological regression grade 6–12 months after initiation of treatment and the clinical prognosis for the next 1–3 years **(Fig. 159).**

During this period clinical progression occurred in 85% of stage T_3 patients who had a locally poor therapeutic effect (R VIII) but in only 13% of such patients with regression grade VI. While the incidence of progression is lower still in patients with regression grades 0–IV, the difference is not statistically significant.

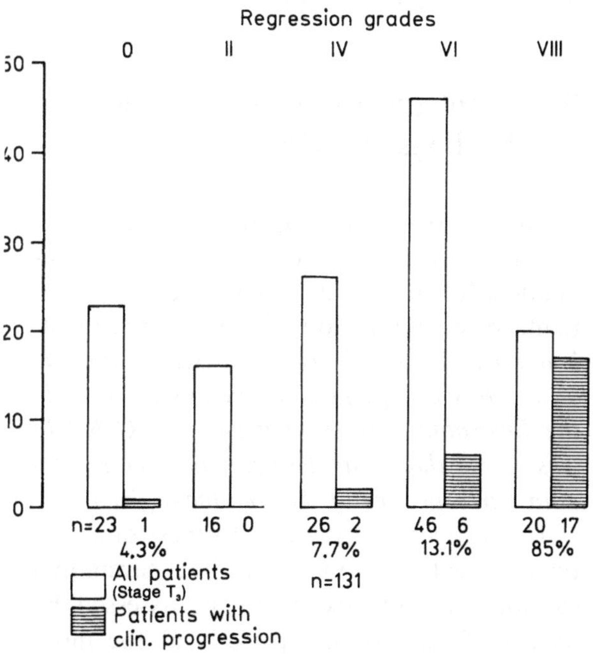

Fig. 159. Relationship between regression grades and clinical progression (observation period 1–3 years)

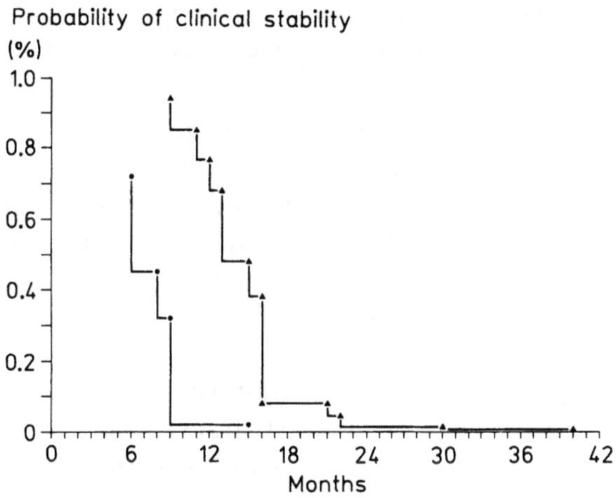

Fig. 160. Correlation between the regression grade as assessed 3 months after commencement of secondary Estracyt/Emcyt therapy and the probability of clinical stability in the further course. ●, patients with an unfavorable regression grade (VIII); ▲, patients with a favorable regression grade (VI). (LEISTENSCHNEIDER and NAGEL 1983)

The validity of cytological regression grading was confirmed in a group of patients who underwent secondary treatment with estramustine phosphate (Estracyt/Emcyt) following proven resistance to hormonal treatment. The average duration of clinical stability was significantly longer in patients exhibiting the favorable regression grade VI 3 months after initiation of the Estracyt/Emcyt treatment than in patients with the poor regression grade VIII (15.5 vs 8.3 months) **(Fig. 160)** (LEISTENSCHNEIDER and NAGEL 1983).

9.6 Signs of Regression After Initiation of Treatment

Signs of regression can often be demonstrated as early as one week after commencement of treatment, especially when this treatment comprises the administration of estrogens or Estracyt/Emcyt (LEISTENSCHNEIDER and NAGEL 1983).

Definitive determination of the regression grade is nevertheless possible only at a later stage, the length of time which must elapse varying according to the form of treatment employed. Thus reliable assessment of the therapeutic effect on the primary tumor can be made at the following intervals after initiation of treatment (LEISTENSCHNEIDER and NAGEL 1980, 1983):

after 3 months: estramustine phosphate (Estracyt/Emcyt)
Cyclophosphamide
5-Fluorouracil

after 6 months: bilateral orchiectomy
estrogens
antiandrogens
LHRH agonists

after 15 months: radiotherapy

When an unfavorable regression grade (R VIII/X) is found after these lengths of time, continuation of the treatment seldom leads to any improvement and therefore resistance to that form of treatment must be assumed. By contrast, favorable regression grades (R IV/VI) may improve still further after continuation of the treatment.

Demonstration of regression grade 0 in quantitatively sufficient and optimally processed cellular material does not mean that the patient is cured of carcinoma, for control by aspiration biopsy has shown that in 79.9% of these patients tumor cells can again be identified subsequently and that their regression grade may even become extremely unfavorable (cytological progression) **(Table 22).** Despite this, clinical progression in the presence of regression grade 0 is on the whole minimal.

If aspiration biopsy nevertheless repeatedly yields regression grade 0, the primary histological or cytological diagnosis should be checked to exclude with certainty an error in the primary diagnosis.

Table 22. Incidence of cytological progression following demonstration of regression grade 0

Findings[a]	*n*	(%)
Primary diagnosis = False diagnosis	4	13.4
Clinical progression	0	0
Cytological progression		
1. Up to regr. grade VI	24	79.9
2. Regr. grade > VI	2	6.7

[a] Observation period 1–4 $^1/_2$ years

In our own case material, repeated cytological determination of regression grade 0 after periods of treatment ranging from 6 months to 5 years (!) led us to have the *primary histological specimens* obtained by punch biopsy re-examined. As a result, the primary histological diagnosis of carcinoma had to be corrected as false-positive in 4 of a total of 30 patients (13.4%) **(Table 22).**

9.7 Cytological Regression Grading and Findings at Palpation

As shown by correlating the cytological regression grade with the evaluation of therapeutic effect by means of palpation in 70 patients receiving various forms of treatment, *in almost 60% of all cases the assessment made on the basis of palpatory findings in cases in the favorable regression grades (R O-R VI) was worse than that derived from cytological data – irrespective of the treatment given (!).*

This proves that the effectiveness of a particular treatment on a prostatic carcinoma cannot even approximately be assessed by means of palpation, especially at the more favorable regression grades. Only when there is poor or no tumoral regression do the findings at palpation correspond to the cytological result **(Fig. 161)** (LEISTENSCHNEIDER and NAGEL 1983).

Consequently it is our opinion that – especially for clinical studies – 'treatment control' by palpation should be rejected as an absolutely unsuitable parameter.

To sum up, it can be stated that for the purpose of primary diagnosis, aspiration biopsy is not only just as good as punch biopsy, it is even superior to it on account of the fact that, unlike punch biopsy, aspiration biopsy is eminently suitable for follow-up investigations during treatment since it causes the patient minimal discomfort and has a very low complication rate, while retaining the diagnostic reliability of punch biopsy.

Furthermore, these advantages show aspiration biopsy to be the only procedure which makes it possible to investigate scientific issues, such as the efficacy of new forms of treatment or problems arising from basic research (see Chap. 13), in a manner acceptable to the patient.

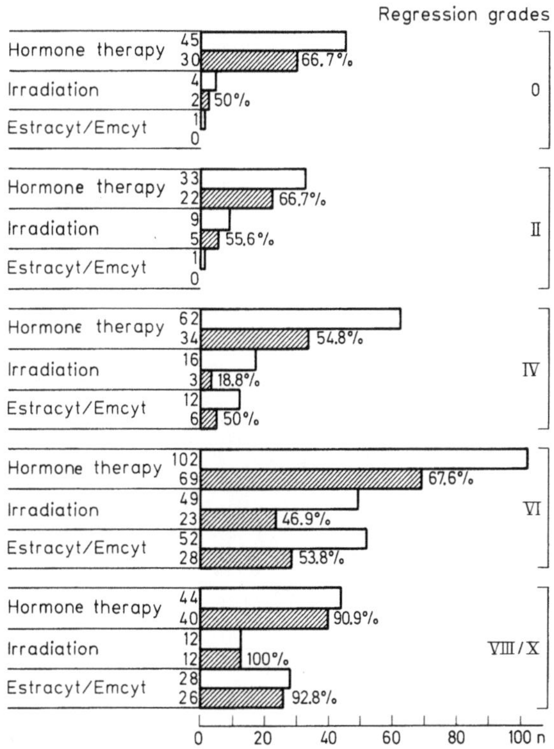

Fig. 161. Correlation between regression grades and palpatory findings typical of carcinoma. Hormone therapy, $n = 286$; irradiation, $n = 90$; Estracyt/Emcyt, $n = 94$

10 Sarcomas

Sarcomas of the prostate are extremely rare, accounting for just 0.1–0.2% of the malignant prostatic tumors (MELICOW et al. 1943; SMITH and DEHNER 1972; TANNENBAUM 1975; NARAYAMA et al. 1978; MÜLLER and WÜNSCH 1981). They occur most frequently during childhood and adolescence, and only 25% of cases present after the age of 40 years. In adults *leiomyosarcoma* is the most frequent lesion; *fibrosarcomas, neurosarcomas* and *osteosarcomas* are observed only rarely and *rhabdomyosarcoma* is practically unknown (SCHUPPLER 1971; MÜLLER and WÜNSCH 1981).

In childhood, by contrast, rhabdomyosarcomas are most common. Myosarcomas account for 65% of *all* prostatic sarcomas, and of these, 60% are rhabdomyosarcomas and 40% leiomyosarcomas. The remaining 35% of lesions are sarcomatous mixed forms (SMITH and DEHNER 1972; KASTENDIECK et al. 1974).

Carcinosarcomas, i.e. tumors with malignant glandular and stromal portions, are still rarer than pure prostatic sarcoma, only three cases having been reported in the literature up to 1981 (KRASTANOVA and ADDONIZZIO 1981). Their histiogenesis has not yet been clarified (WAJSMAN and MOTT 1978).

Prostatic sarcoma can be reliably distinguished from carcinoma using cytological techniques.

Even at the low screening magnification, complete dissociation of the tumor cells can be recognized **(Fig. 162a, e).** The *nuclei* are extremely large and considerably polymorphic. Along with round nuclei, round to oval and oval nuclei dominate; occasional wedge-shaped nuclei are also found **(Fig. 162b–d).** The cells often display no cytoplasm and are designated 'naked nuclei' tumor cells **(Fig. 162e, f).** The *chromatin structure* is very condensed, in part with clumping **(Fig. 162c, d).**

The nucleoli are prominent, with evident loss of circularity **(Fig. 162d).** In addition to hyperchromatic nuclei one also sees hypochromatic nuclei containing nuclear inclusions (MÜLLER and WÜNSCH 1981). In addition, *local perinuclear haloes (lighter areas of cytoplasm)* also occur **(Fig. 162d).**

Histogenetic classification of the sarcomas is not possible using cytological methods alone. Only the extremely rare osteosarcoma can be precisely diagnosed histogenetically, this on the basis of the minute osseous fragments seen in the cytological smear (MÜLLER and WÜNSCH 1981).

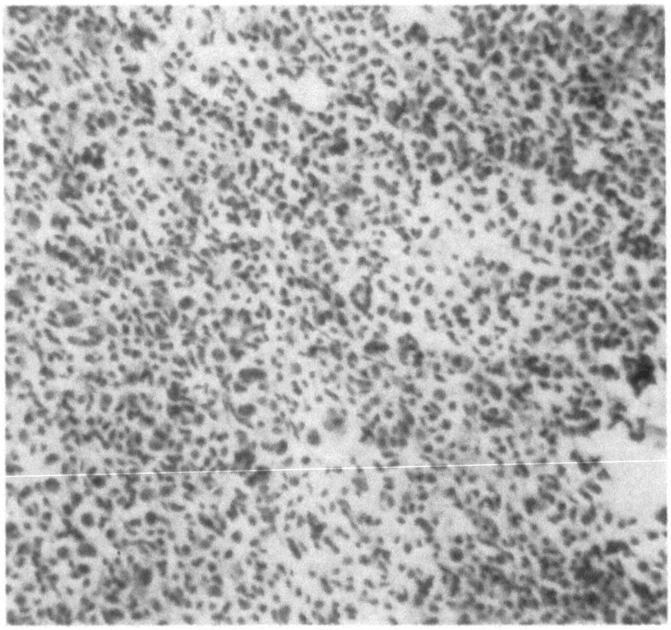

Fig. 162a. Prostatic sarcoma at the low magnification used for preliminary screening, with the typical complete nuclear dissociation, ×100

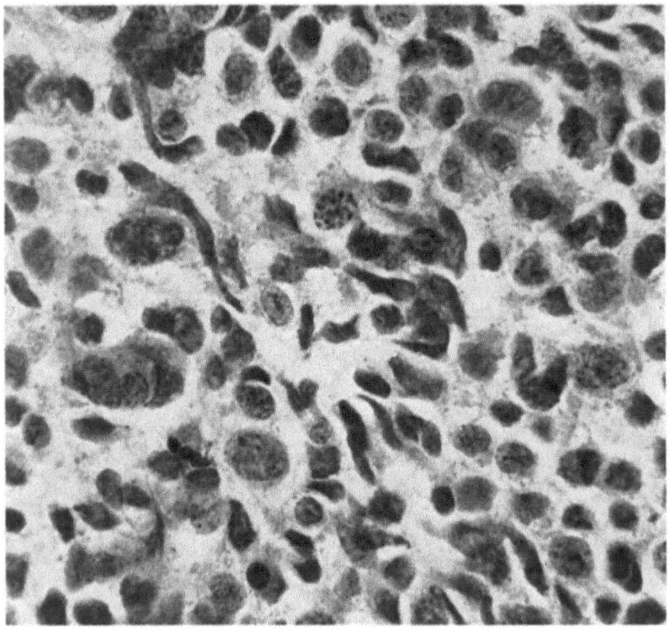

Fig. 162b. Same case at higher magnification: considerable nuclear polymorphism; cell borders scarcely distinguishable or totally lost. ×400

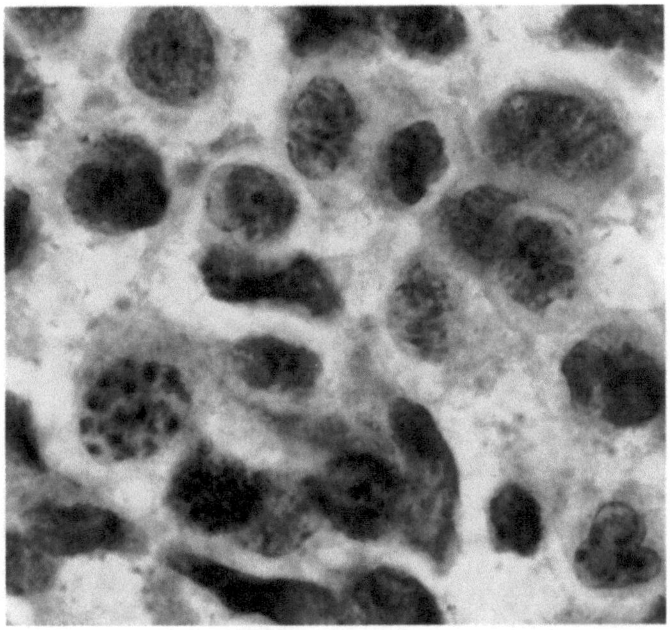

Fig. 162c. Same case at the highest magnification: mainly oval nuclei are seen alongside round ones, though some are polygonal or wedge-shaped. The nuclear chromatin structure is coarsened; there is some clumping, and abundant chromatin aggregates are present locally. The nucleoli are relatively small. Oil immersion, ×1000

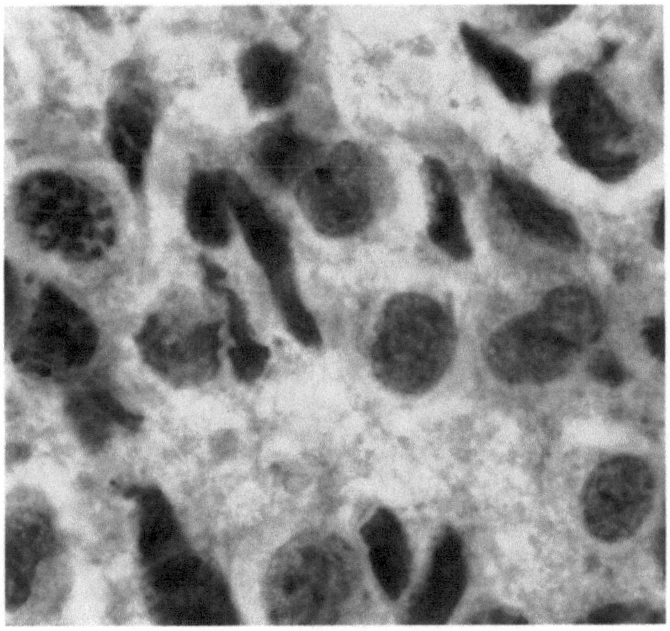

Fig. 162d. Same case, again at highest magnification: numerous wedge- to spindle-shaped nuclei. Relatively small nucleoli. Isolated areas of pale cytoplasm perinuclearly (perinuclear haloes). Oil immersion, ×1000

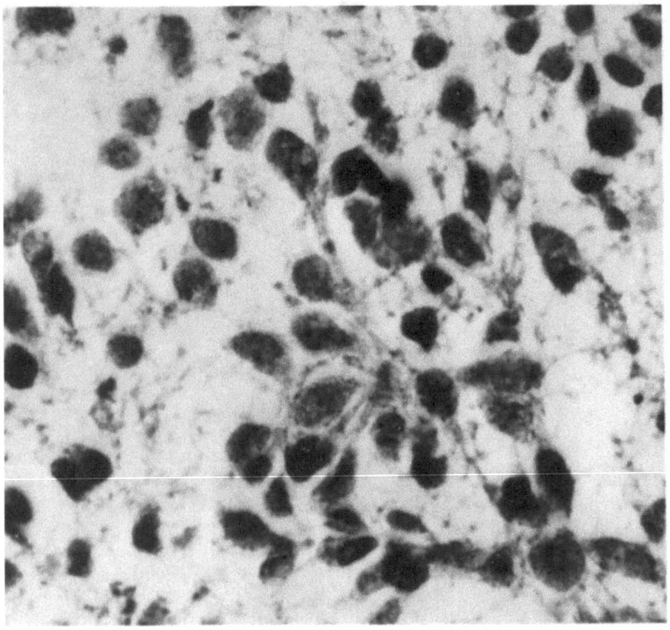

Fig. 162e. Same case, with still more pronounced nuclear dissociation. Numerous 'naked nuclei' tumor cells. ×400

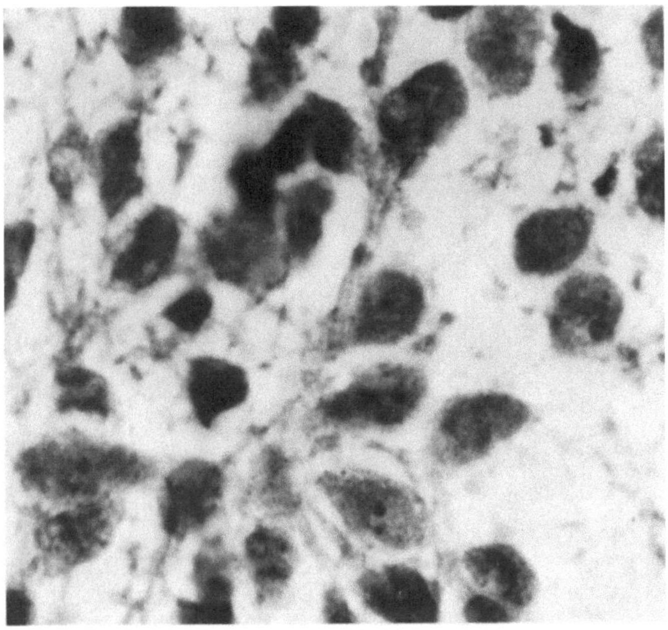

Fig. 162f. Same case at higher magnification: bizarre nuclear shapes; on the whole, complete loss of the nuclear membranes. Some nucleoli here are more prominent. ×630

11 Secondary Tumors of the Prostate

Secondary tumors of the prostate are benign or malignant tumors of extraprostatic origin which gain access to the prostate from other organs by way of lymphatogenous or hematogenous spread or direct infiltration.

Direct infiltration, i.e. per continuitatem, results above all from carcinomas of the urinary bladder or large intestine.

Hematogenous or lymphatogenous metastasis to the prostate occurs particularly from malignant lymphomas, bronchial carcinomas, gastrointestinal carcinomas, penile carcinomas, laryngeal carcinomas, malignant tumors of the testes, plasmacytomas and melanomas (SMITH and DEHNER 1972; JOHNSON et al. 1974; MELCHIOR et al. 1974; CARTAGENA and BAUMGARTNER 1975; DAJANI and BURKE 1976; WAJSMAN and MOTT 1978).

Of all malignant tumors of the prostate, 1.2% are metastatic secondary tumors (JOHNSON et al. 1974) **(Table 23).**

The frequency of leukemic infiltration of the prostate is not clear. While MELCHIOR et al. (1974) found secondary prostatic involvement in only 1.7% of autopsies on patients with systemic leukemia, they cited the frequency as 22.5% after specific surgical procedures.

Carcinoma of the bladder usually results in infiltration of the prostate only at an advanced stage. Neither the incidence of such infiltration nor that of infiltration by carcinomas of the large intestine has yet been clarified.

In our own case material, the proportion of secondary prostatic tumors found among 1,086 aspiration biopsies at primary diagnosis was 1.1% **(Table 24),** while in the same biopsy material primary prostatic carcinoma was diagnosed in 29% **(Table 7).**

11.1 Cytomorphological Criteria

11.1.1 Urothelial Carcinoma

Secondary urothelial carcinomas of the prostate almost exclusively arise through direct infiltration from the bladder. In most cases far advanced bladder carcinoma is present. According to the TNM classification (UICC 1979) these are T_{4a} cases; at this stage lymph node metastases can usually be identified and distant metastases are often found, too. The grade of malignancy is correspondingly high, the carcinomas chiefly being assigned to grades II and III.

As a result, cytological diagnosis poses no problems, particularly when abundant prostatic epithelium is present in the same smear.

The nuclei are large, predominantly oval and mostly distinctly polymorphic. The chromatin is very condensed, irregularly distributed and aggregated in places **(Figs. 163, 164).**

The nucleoli are very prominent and in part polymorphic. Most of the nuclei contain two or three nucleoli **(Figs. 168, 169).** The nuclear pattern is always disturbed. The cell clusters show partial dissociation and uni- or multinucleated giant cells are not infrequent **(Fig. 164).** Of importance for differential di-

Table 23. Incidence of secondary metastatic tumors of the prostate in the presence of various malignant primary tumors (JOHNSON et al. 1974).

Primary tumor	Histological type	*n*	Prostatic involvement	
			n	(%)
Skin	Malignant melanoma	1779	9	0.6
Lungs	Bronchial carcinoma	4323	5	0.2
Pancreas	Adenocarcinoma	273	1	0.7
Stomach	Adenocarcinoma	605	1	0.2
Penis	Squamous cell carcinoma	220	1	0.4
Larynx	Adenocarcinoma	1286	1	0.1

Table 24. Nature and distribution of secondary prostatic tumors in 1086 aspiration biopsies according to Franzén

Tumor	*n*	%
Urothelial carcinoma	8	0.7
Malignant lymphoma	3	0.3
Seminoma	1	0.09

agnosis are the typical elongated, often tail-like cytoplasmic extensions and the abundant development of any remaining cytoplasm **(Figs. 165, 168, 169).** Fine or somewhat more coarse vacuolation and perinuclear haloes are often seen in the cytoplasm **(Figs. 165, 168).**

11.1.2 Malignant Lymphoma

Malignant lymphoma and chronic lymphatic leukemia are probably different terms for the same basic disease; they *cannot* be distinguished cytologically.

It is equally impossible to differentiate secondary and primary malignant lymphomas of the prostate cytologically, but the latter are considerably less common.

Cytomorphologically there is always an abundance of cells. Since it is generally the case that the entire prostate is involved, prostatic epithelial cells are only very occasional findings. The picture is in fact dominated by often totally dissociated naked nuclei of varying size. These nuclei are strikingly round and only moderately polymorphic **(Fig. 170a, b).** The nuclear membranes are sharply defined, and some are indented **(Figs. 170b, c, 171).** The chromatin pattern may be finely granular and reticular, but it can also be distinctly condensed and coarsened. The nucleoli are moderately prominent, and most nuclei contain at least two **(Fig. 170b, c).**

11.1.3 Seminoma

Hematogenous or lymphatogenous metastasis to the prostate from testicular tumors is extremely rare. Like all epithelial tumors, seminomas can be well, moderately well or poorly differentiated (THACKRAY and CRANE 1976).

At least minor accompanying lymphocytic infiltration can be demonstrated in 90% of cases, and in 25% this infiltration is substantial. Thirty per cent of cases show a granulomatous reaction with lymphocytes, plasma cells, epithelioid cells and Langhans' giant cells. In 10% of the cases this reaction is considerable, and in 6–11% syncytial giant cells can be demonstrated.

The rare spermatocytic seminomas, which have an age-dependent incidence of between 1.9% and 9% (THACKRAY and CRANE 1976; MOSTOFI and PRICE 1977; HEDINGER 1978), represent a special tumor form. The cytological picture of seminoma is consequently very variable.

Cytologically, well differentiated seminoma is characterized by uniform, medium-sized, round nuclei. On account of its typical glycogen and lipid content, the cytoplasm is predominantly clear or finely to coarsely vacuolated. The cell borders are easily distinguishable.

With increasing dedifferentiation the nuclei become exceedingly large and polymorphic, with a polygonal or spindle-shaped appearance. Most of the nuclear membranes are no longer visible **(Fig. 172e, f)**. The chromatin is very condensed, and the number of nucleoli per nucleus ranges from two to four. The nucleoli are prominent and polymorphic **(Fig. 172c–f)**. The size of the nuclei corresponds to that in grade III prostatic carcinoma.

Essential criteria for differential diagnosis from prostatic carcinoma are the large number of mitoses (one or two per microscopic field, **see Fig. 172b–d***) and the regular occurrence of karyorrhexis* **(Fig. 172c, f)** *in poorly differentiated seminomas.*

Seminomas with a lymphocytic or granulomatous reaction can be reliably classified cytologically.

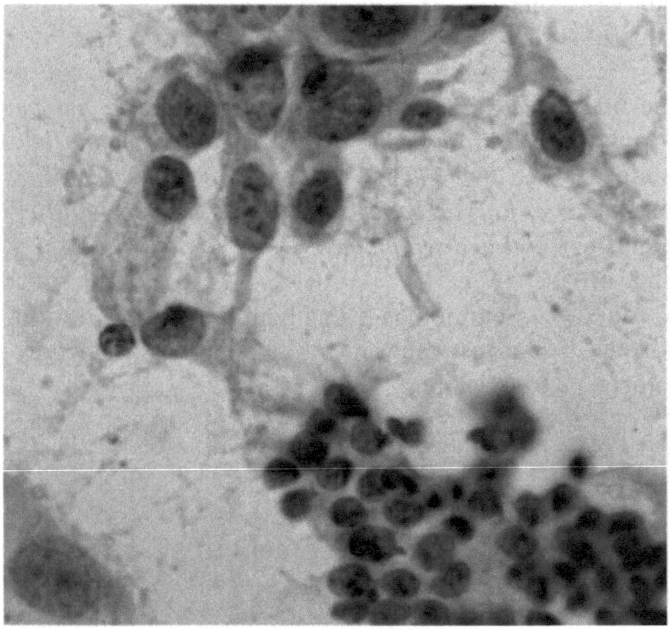

Fig. 163. Group of urothelial carcinoma cells (*upper part of the picture*) alongside normal prostatic epithelium (*below*). Strikingly large, oval, moderately polymorphic nuclei of urothelial carcinoma. Dense, relatively homogeneous chromatin structure, with localized clumping. ×630

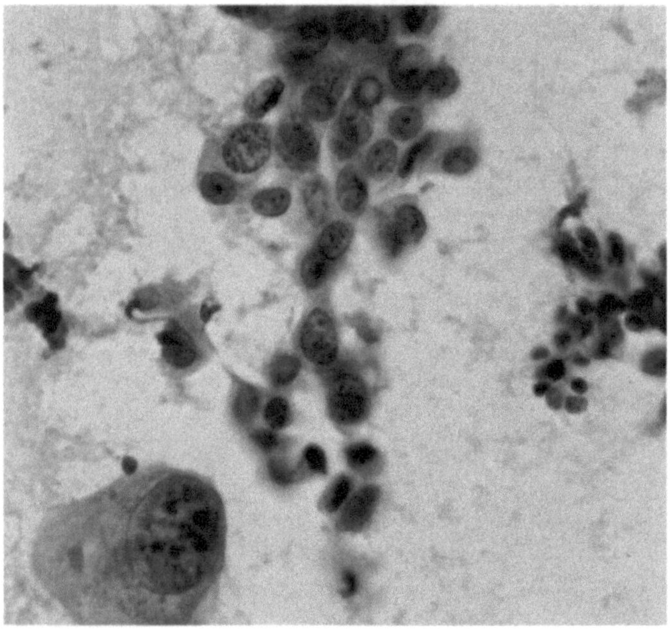

Fig. 164. Group of urothelial carcinoma cells and an isolated tumor giant cell with chromatin aggregates. *Right,* a small sheet of normal prostatic epithelium. ×400

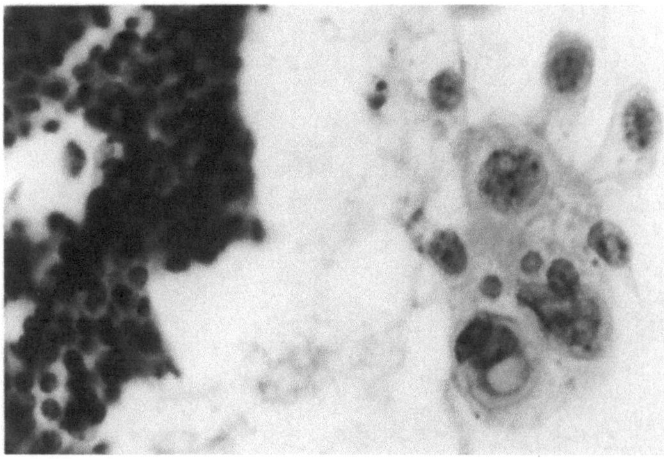

Fig. 165. Normal prostatic epithelium (*left of picture*) and small group of urothelial carcinoma cells (*right*) with varying degrees of cytoplasmic vacuolation. Localized perinuclear haloes. ×400

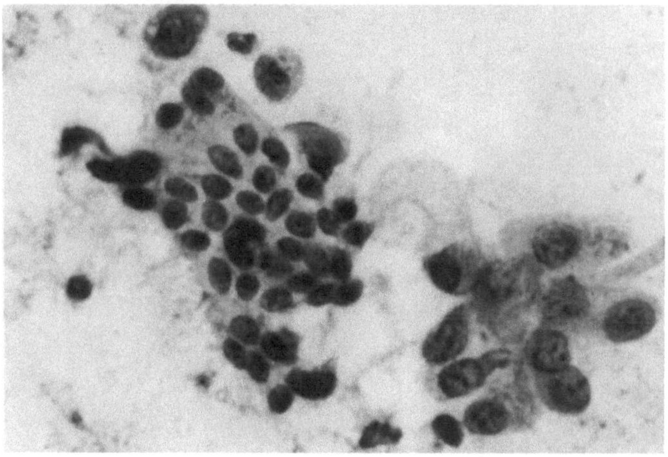

Fig. 166. Small sheet of normal prostatic epithelium with two overlying urothelial carcinoma cells, and a group of urothelial carcinoma cells displaying prominent and polymorphic nucleoli. ×400

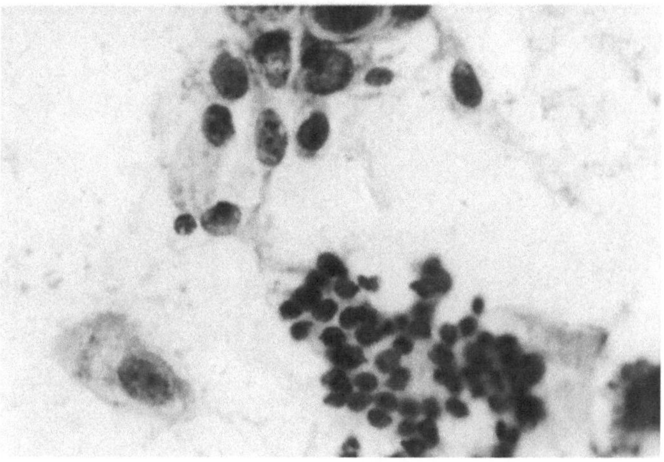

Fig. 167. Group of urothelial carcinoma cells (*top*) adjacent to a sheet of normal prostatic epithelium. ×400

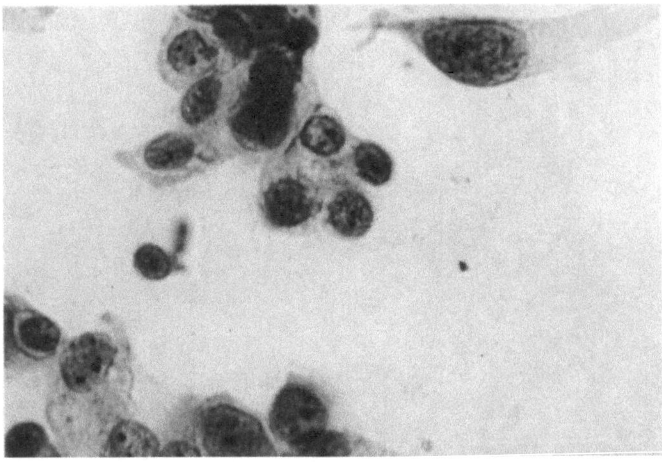

Fig. 168. Small group of urothelial carcinoma cells with a clearly recognizable, typically eccentric nuclear location and abundant cytoplasm locally. Tail-like cytoplasmic extensions of the isolated carcinoma cell (*top right*). ×400

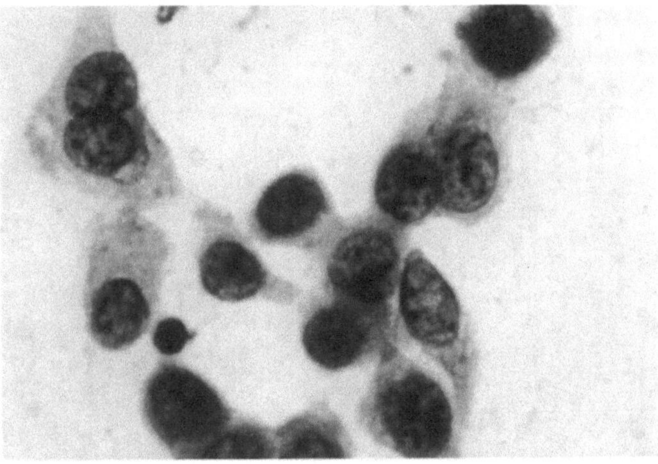

Fig. 169. Group of urothelial carcinoma cells at higher magnification: localised tail-like cytoplasmic extensions, eccentrically located nuclei, more than one nucleolus per nucleus, marked nuclear polymorphism. ×630

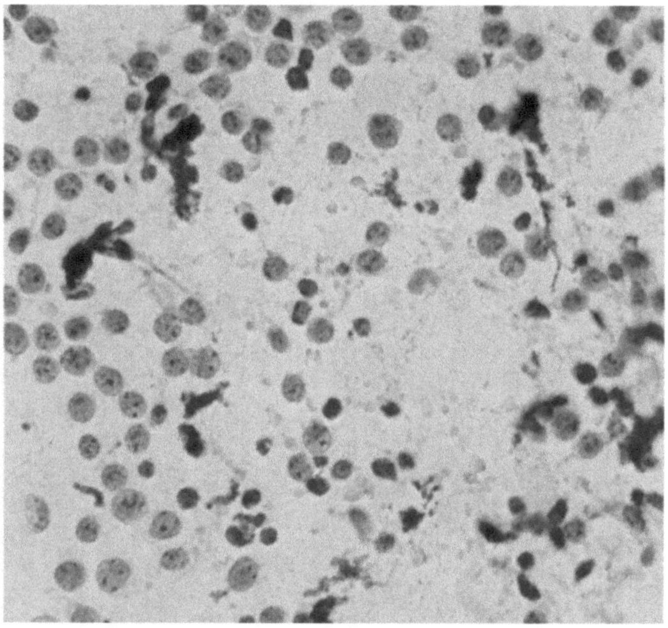

Fig. 170a. Cells from a malignant lymphoma in the prostate. Totally dissociated, moderately polymorphic, relatively small tumor cells. × 400

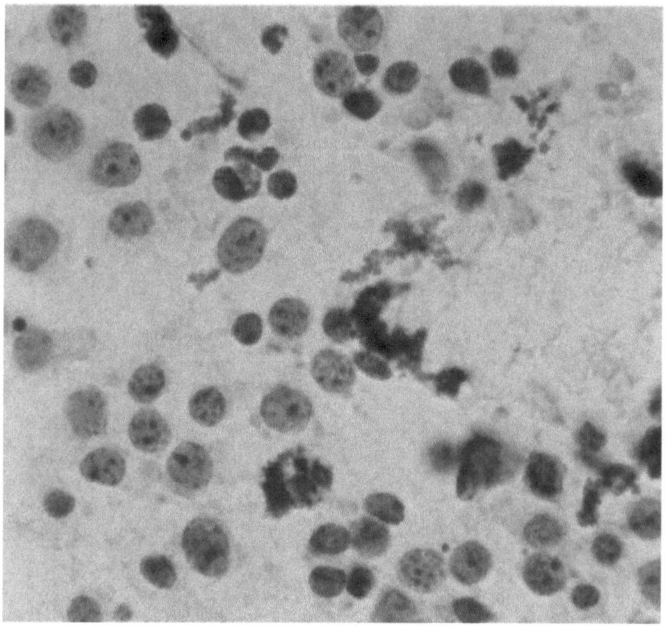

Fig. 170b. Section from the same smear at higher magnification: the lymphoma cells display easily recognizable, clearly outlined nuclear membranes. Finely granular chromatin structure and moderately prominent, predominantly polymorphic nucleoli; in most cases two or more nucleoli per nucleus. × 630

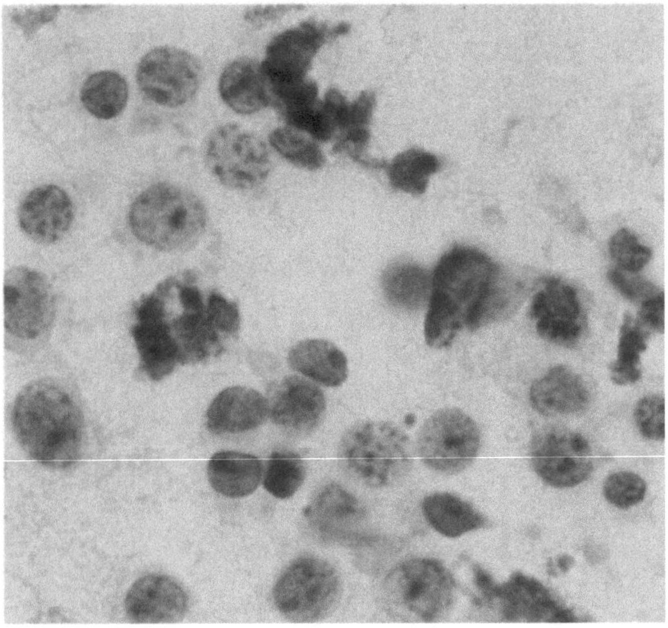

Fig. 170c. Same case at the highest magnification: good demonstration of the typical finely granular, loose chromatin structure. Oil immersion, ×1000

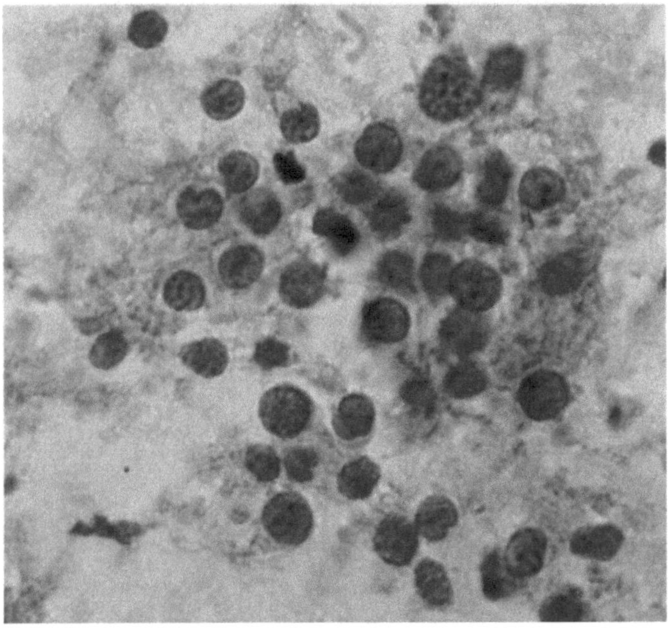

Fig. 171. Another case of malignant lymphoma in the prostate. Here the nuclei are rather less dissociated. Clearly discernible, localized, typical indentations of the nuclear membranes. ×630

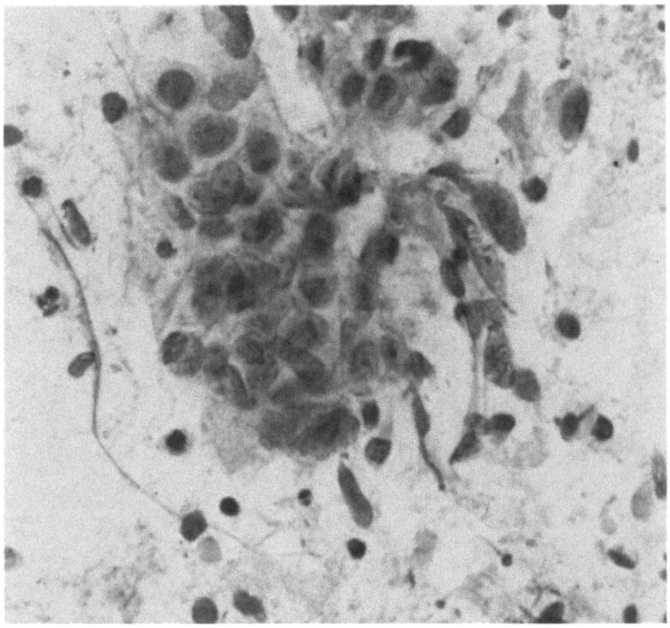

Fig. 172a. Group of cells from a seminoma in the prostate. ×400

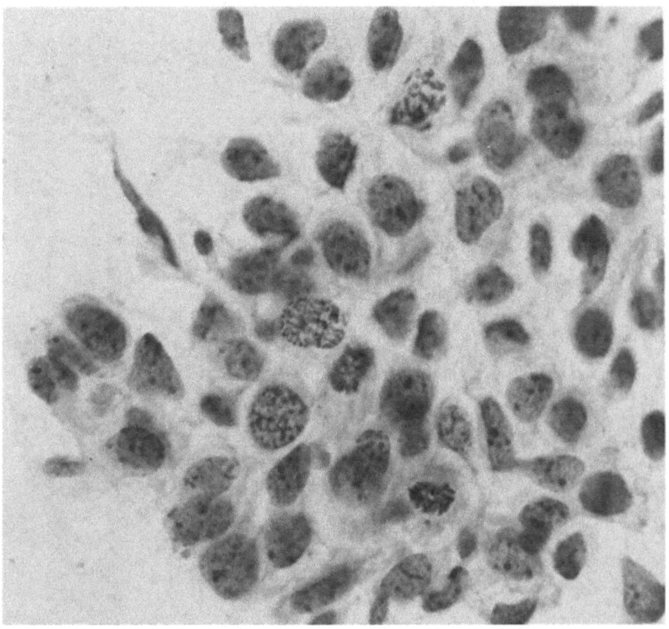

Fig. 172b. Same case, a different aggregate: high-grade nuclear polymorphism; very condensed chromatin, often with clumping; cells in mitosis. ×400

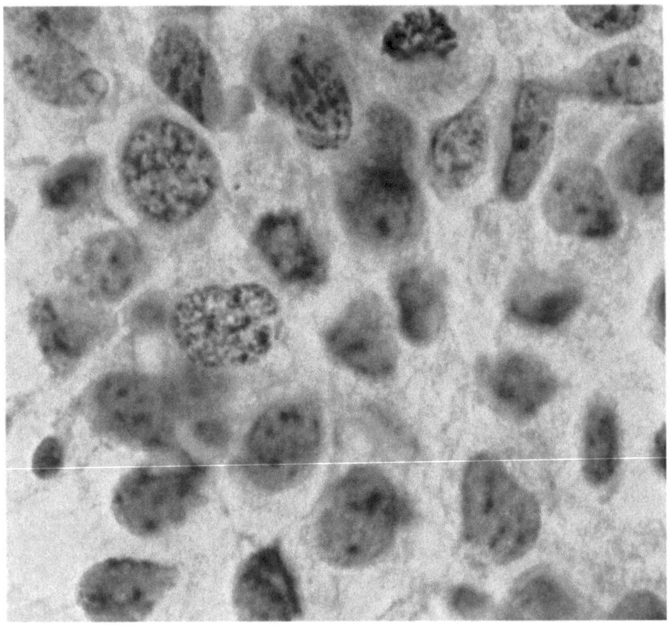

Fig. 172c. Same aggregate at higher magnification: numerous polygonal, in part wedge-shaped nuclei, with several nucleoli per nucleus. A tumor cell in mitosis is seen at the *upper edge of the picture;* isolated karyorrhectic cells. ×630

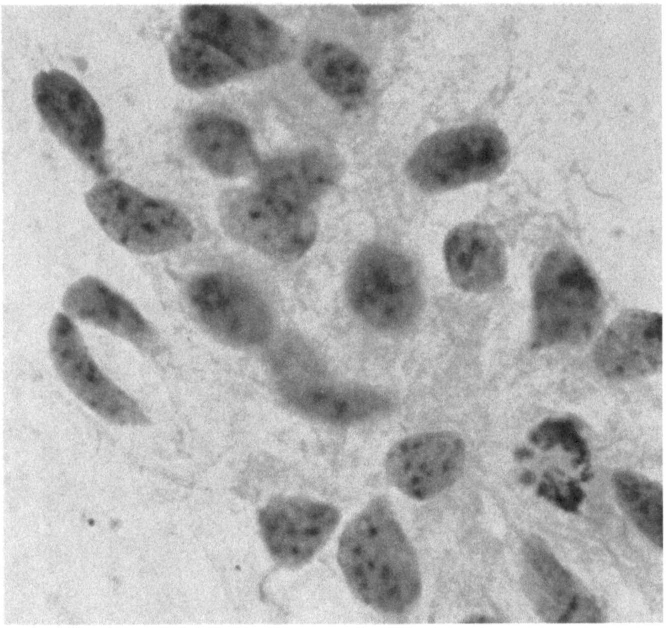

Fig. 172d. Another group of cells from the same case; again a cell undergoing mitosis is seen. ×630

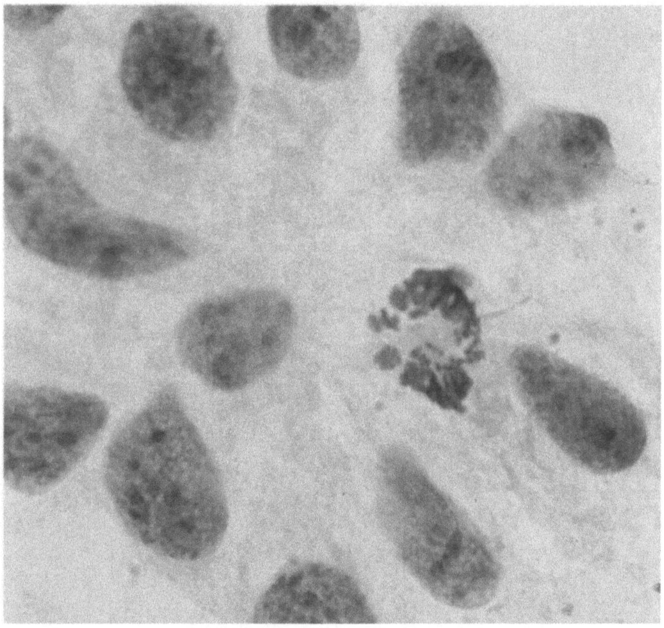

Fig. 172e. Same group at the highest magnification: in addition to the typical, very dense chromatin structure and the increased number of nucleoli, the loss of the nuclear membranes is evident. Oil immersion, × 1 000

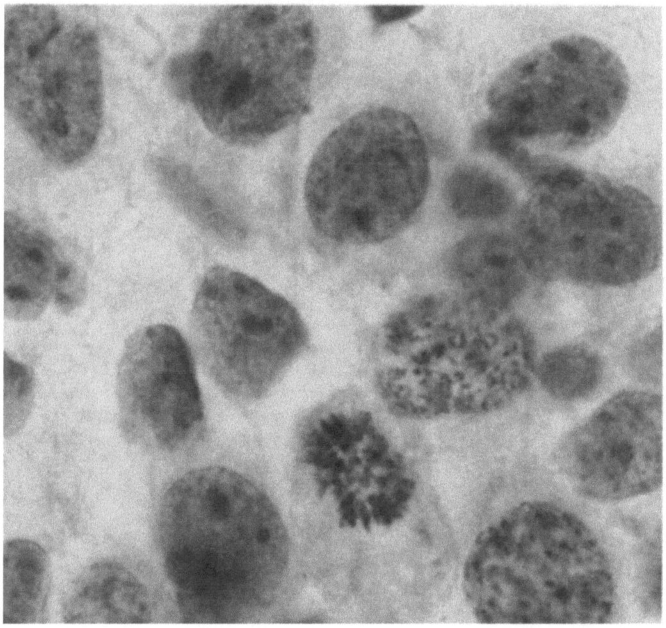

Fig. 172f. Another group of cells at the highest magnification, with localized karyorrhexis and yet another cell in mitosis. Oil immersion, × 1 000

12 Prostatitis

Both nonspecific and specific prostatitis may be focal or diffuse in character, and, correspondingly, varying numbers of inflammatory cells are demonstrated in aspirates.

Focal prostatitis may take an intraductal, periductal or interstitial form (NILSSON et al. 1973; LEISTENSCHNEIDER and NAGEL 1978; KOHNEN and DRACH 1979).

12.1 Classification

The *histological classification* of prostatitis distinguishes six forms **(Table 25).**

The *cytological* diagnosis of prostatitis is just as reliable as histological diagnosis (ZIEGLER and VÖLTER 1975; ESPOSTI et al. 1975; FAUL 1975); however, the subclassification into *periductal* and *interstitial* forms, permitted by histological investigation, is not possible with cytological methods since no stroma can be obtained by means of aspiration biopsy.

Aspiration does nevertheless permit, on the basis of cytological criteria, distinction between the following six types of prostatitis – corresponding to the histological classification cited above (LEISTENSCHNEIDER and NAGEL 1978, 1979):

purulent prostatitis
abscessing prostatitis
chronic prostatitis
chronic recurrent prostatitis
granulomatous prostatitis
tuberculous prostatitis

The most frequent type is chronic recurrent prostatitis while tuberculous prostatitis is the rarest **(Fig. 173).**

Table 25. Classification and frequency of histologically proven cases of prostatitis in 1000 punch biopsies (LEISTENSCHNEIDER and NAGEL 1978)

Type of prostatitis	*n* (101)	%
Purulent prostatitis	4	3.96
Abscessing prostatitis	3	2.97
Chronic prostatitis		
a) Periductal form	40	39.60
b) Interstitial form	25	24.75
Chronic recurrent prostatitis	12	11.88
Granulomatous prostatitis	13	12.87
Tuberculous prostatitis	4	3.96

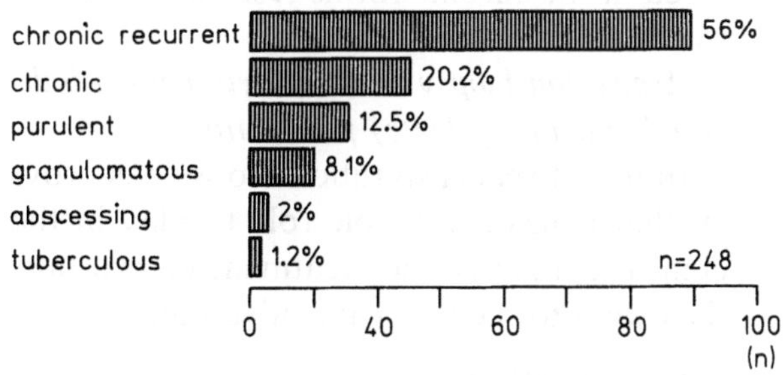

Fig. 173. Frequency of the various forms of prostatitis in our own biopsy material (LEISTENSCHNEIDER and NAGEL 1981)

12.2 Diagnostic Reliability

In 124 simultaneously performed aspiration and punch biopsies, there was agreement between the diagnoses and classifications of prostatitis obtained by the cytological and histological methods respectively in 90.1% of cases (LEISTENSCHNEIDER and NAGEL 1981) **(Table 26).**

Table 26. Rate of agreement between cytological and histological diagnosis of prostatitis (LEISTENSCHNEIDER and NAGEL 1981)

Type of prostatitis	Histology *n*	Cytology *n*
Purulent	11	11
Abscessing	3	3
Chronic	43	36
Chronic recurrent	49	46
Granulomatous	14	13
Tuberculous	4	3
	124	112 (90.3%)

12.3 Clinical Significance and Complications

Aspiration biopsy – should it be necessary – is the last of all the clinical measures involved in the confirmation and/or differential diagnosis of prostatitis. It is especially indicated in the chronic forms resistant to therapy.

Aspiration biopsy is absolutely contra-indicated in acute (febrile) prostatitis.

In recent years aspiration biopsy has proven that it has a valuable role to play in the clinical diagnosis and treatment of prostatitis, owing to the following advantages:

- low complication rate
- classification of the diverse forms of prostatitis.
- diagnosis of tuberculous prostatitis
- differential diagnosis between prostatitis and carcinoma
- possibility of aiding the planning of treatment and, when appropriate, of functioning as a means of treatment control.

As indicated in this list, aspiration biopsy is of the utmost significance not only for the cytological diagnosis and classification of prostatitis but also for differential diagnosis between prostatitis and carcinoma, which is often difficult by palpation alone.

The clinical relevance of this diagnostic problem is illustrated by the results of recent investigations:

in 10–30% of aspirations performed because of palpatory findings suggestive of carcinoma, cytological examination revealed not the suspected carcinoma but considerable inflammation (ESPOSTI et al. 1975; FAUL 1975; LEISTENSCHNEIDER and NAGEL 1978, 1981).

On the other hand, aspiration biopsy disclosed prostatic carcinoma in approximately 10% of patients with clinically diagnosed prostatitis (ESPOSTI et al. 1975).

12.4 General Cytological Criteria of Prostatitis

It is characteristic of all forms of prostatitis that numerous inflammatory cells are present and that there are atypias of varying degrees (Pap II–IV).

The large number of inflammatory cells, which dominate the picture at the preliminary examination at low magnification, is pathognomonic for prostatitis; the details, however, can be assessed only at higher magnification. These inflammatory cells often infiltrate the sheets of prostatic epithelium, though they may be absent from certain areas of the smear at higher levels of magnification. The atypias are mostly only low grade (Pap

II) in cases of purulent or abscessing prostatitis (Fig. 174); in chronic, chronic recurrent and granulomatous prostatitis, by contrast, medium-grade atypias (Pap III) predominate **(Figs. 191, 192, 212, 213),** and even high-grade atypias (Pap IV) may be present **(Fig. 214).** In tuberculous prostatitis the atypias are invariably of the Pap IV type **(Fig. 226).**

Occasionally the presence of very pronounced atypias in an aspirate may render distinction from a well-differentiated carcinoma (G I) so difficult that after appropriate anti-inflammatory treatment control biopsy is essential (FAUL 1975; LEISTENSCHNEIDER and NAGEL 1981).

If the findings are still ambiguous, punch biopsy should be performed before finally excluding carcinoma, as this method permits evaluation not only of cells and nuclei but also of the stroma and thus affords additional information which may help in reaching a definitive diagnosis.

Prostatitis can be reliably diagnosed and classified cytologically on the basis of (a) the patterns of inflammatory cells characteristic for the individual forms and (b) the cellular atypias that can always be demonstrated **(Table 27).**

Table 27. Inflammatory cell pattern and severity of epithelial atypias in the various forms of prostatitis (LEISTENSCHNEIDER and NAGEL 1981)

Type of prostatitis	Inflammatory cell pattern	Epithelial atypias		
		Mild	Moderate	Severe
Purulent	Leukocytes – diffuse	+	–	–
Abscessing	Leukocytes – diffuse – clusters	+	–	–
Chronic	Lymphocytes Histiocytes (GC)[a] Plasma cells	+	+	–
Chronic recurrent	Leukocytes Lymphocytes Plasma cells Histiocytes (GC)[a]	–	+	+
Granulomatous	Leukocytes (eosinophilic) Lymphocytes Plasma cells Histiocytes – GC – clusters Langhans' – GC	–	+	+
Tuberculous	Leukocytes Lymphocytes Plasma cells Histiocytes – clusters Langhans' – GC Epithelioid cells – diffuse – clusters	–	–	+

[a] *GC*, giant cells

12.4.1 Specific Forms

12.4.1.1 Purulent Prostatitis (Figs. 174–179)

Cell borders are often indistinct. The sheets of epithelium vary in size, are sometimes dissociated, and are more or less diffusely infiltrated by polymorphonuclear leukocytes, which are also scattered throughout the entire smear. The epithelial atypias are low-grade. The nuclear chromatin structure is predominantly loose; there are also local anisokaryosis and isolated prominent nucleoli. Overall, however, the atypias correspond only to Pap II.

12.4.1.2 Abscessing Prostatitis (Figs. 180, 181)

Typically, there is a fibrinous-leukocytic background to the smear. Equally striking are the massive clusters of leukocytes. The cell sheets are often distinctly dissociated and infiltrated by leukocytes; the cell borders are no longer detectable.

The epithelial atypias are the same as those seen in purulent prostatitis (Pap II).

Although suppurative and acute purulent prostatitis display similarities cytologically, they can be distinguished unequivocally on the basis of the massive leukocytic clustering present in the form with abscesses.

12.4.1.3 Chronic Prostatitis (Fig. 182)

Typical of this form is an inflammatory cell pattern which is *without* leukocytes and comprises only loose collections of round cells and strikingly few plasma cells and histiocytes. When histiocytes can be demonstrated, they usually display finely vacuolated, abundant cytoplasm. Only some of the sheets of epithelium are infiltrated by such inflammatory cells; the latter are mainly located in the direct proximity of these sheets and overall are only loosely scattered over the entire smear. The sheets of epithelium are generally well preserved or only slightly dissociated. The cell borders are on the whole intact. Nuclei are small to medium sized, and the chromatin structure is loosely granular. Occasionally, conspicuous nuclei are present.

The atypias correspond to Pap II–III.

12.4.1.4 Chronic Recurrent Prostatitis (Figs. 183–192)

The very typical variegated picture presented by the inflammatory cells comprises a quite uniform mixture of round cells, histiocytes and above all – in contrast to the situation in chronic prostatitis – numerous leukocytes. Most of the histiocytes display more pronounced cytoplasmic changes, with marked, fine to coarse vacuolation; the cytoplasm also frequently has a foamy, granular appearance. It is not unusual for the histiocytes to contain two or more nuclei.

There is a varying degree of infiltration of prostatic epithelium by the inflammatory cells. Alongside larger sheets of epithelium, many smaller sheets are seen; depending upon the severity of the inflammation, they show differing degrees of dissociation and blurring of cell borders.

In this form of inflammation it is characteristic to find medium-grade epithelial atypias with mild nuclear polymorphism, a slightly condensed chromatin structure and locally conspicuous nucleoli. The nuclear arrangement may be moderately disturbed **(Figs. 191, 192).**

Since dense chromatin and prominent nucleoli are never encountered, differential diagnosis from prostatic carcinoma is not difficult.

The dissemination of the inflammatory cells within the epithelial sheets varies according to the severity of the inflammation, which must always be recorded in the evaluation of any case of prostatitis.

Prostatic Infarction (Figs. 193–197b)

In prostatic infarction the cytological picture is basically the same as in chronic recurrent prostatitis. Nevertheless, among the inflammatory cells there is an extremely high proportion of histiocytes of a most diverse form and size. Most of them are multinucleated; their cytoplasm is generally moderately to coarsely vacuolated. Loosely scattered squamous cells resulting from the metaplastic transformation of prostatic epithelium which takes place in the area of infarction are also a regular feature.

12.4.1.5 Granulomatous Prostatitis (Figs. 198–214)

As with chronic recurrent prostatitis, numerous inflammatory cells are seen in a variegated pattern. In special forms, eosinophilic granulocytes are detected in addition to the normal polymorphonuclear leukocytes. Plasma cells, on the other hand, are more common in chronic recurrent prostatitis.

The following may be regarded as reliable criteria for differentiating from chronic recurrent prostatitis:

- The regular occurrence of histiocytes containing more than one nucleus; often these histiocytes actually contain many nuclei or occur as exceedingly large histiocytic giant cells **(Fig. 205).**

- The cytoplasm often has a foamy, granular, or moderately to coarsely vacuolated structure; not infrequently it contains phagocytized nuclear fragments. Occasionally Langhans' giant cells are also present **(Figs. 198, 209, 210).**
  Foreign body giant cells occasionally become confluent, forming multinucleated giant cells, and through a corresponding arrangement may mimic so-called micro-adenomas. These giant cells with peripherally located nuclei must be differentiated from the latter, which are observed in carcinoma (Wullstein and Müller 1973).
  The decisive criterion for this differentiation is the nuclear structure, which, in histiocytic giant cells, is uniformly oval and displays a loose chromatin pattern. Nucleoli, which are also of importance for differential diagnosis, are at most 'conspicuous' in these cells, never 'prominent'!

- Along with single histiocytes, characteristic clusters of histiocytes are to be observed, occasionally with transitional forms into epithelioid cells **(Figs. 200, 202, 203, 206)** (Leistenschneider and Nagel 1978, 1979).

- The sheets of epithelium resemble those of chronic recurrent prostatitis structurally; some are well preserved, others markedly dissociated **(Figs. 203, 212).**

- Overall, the epithelial atypias are more pronounced and the nuclei of varying size, so that they are sometimes as large as those of grade I carcinomas. In addition the chromatin is condensed and the nuclear arrangement often considerably disturbed. The nuclear membranes are not always recognizable. The nucleoli are mostly only conspicuous, but in some areas of the epithelial sheets they may be prominent **(Figs. 212–214).**

The lack of regular demonstration of prominent nucleoli and the simultaneous abundance of inflammatory cells generally allow definite differentiation from prostatic carcinoma, the epithelial atypias being only of the Pap III type or occasionally III to IV.

Nevertheless there are cases in which the atypias present in granulomatous prostatitis correspond to Pap IV findings, so that the differential diagnosis from carcinoma can be just as problematic cytologically as that from a solid anaplastic carcinoma is histologically (Hohbach et al. 1980).

When differentiating between granulomatous prostatitis and prostatic carcinoma it must also be borne in mind that in 10% of cases of granulomatous prostatitis, a prostatic carcinoma is found simultaneously (Hohbach et al. 1980).

12.4.1.6 Tuberculous Prostatitis (Figs. 215–226)

Tuberculous prostatitis can be reliably differentiated from granulomatous prostatitis by cytological techniques. The following characteristics are decisive in this distinction:

- Large collections of histiocytes and epithelioid cells, mostly in the form of clusters. These cell clusters are more pronounced than in granulomatous prostatitis **(Figs. 216, 218, 220–225).**
- Dirty background to the entire smear, and large areas of amorphous, grey to grey-pink, partially granular material, which stems from the necrosis typical of tuberculosis **(Figs. 215–219).**
- Severe epithelial atypia (Pap IV) with markedly enlarged nuclei and anisokaryosis, condensed nuclear chromatin, conspicuous to prominent nucleoli, and mainly still distinguishable nuclear membranes **(Figs. 214, 226).**

Co-existence of epithelioid cell clusters, sheets of prostatic epithelium and necrotic material is typical **(Figs. 216, 220).** The sheets of epithelium themselves are in part still intact, in part extensively dissociated at their margins and to varying degrees infiltrated by variegated inflammatory cells, as in chronic recurrent prostatitis **(Figs. 221, 224).**

The atypias demonstrated in tuberculous prostatitis correspond at least to Pap III findings, and locally even to Pap IV, making differentiation from carcinoma occasionally difficult.

12.5 Summary

Six forms of prostatitis may be distinguished cytologically.

Typical for the individual forms is the severity of the cellular atypias in combination with the presence of diverse inflammatory cells.

Prostatitis may be diagnosed cytologically only when ample material is available and the inflammatory changes are marked. Isolated inflammatory cells seen outside of the sheets of prostatic epithelium do not suffice for the diagnosis of prostatitis.

Especially the differentiation of granulomatous and tuberculous prostatitis from carcinoma is difficult on occasion, and more pronounced chronic recurrent prostatitis can also cause problems in this respect.

Before the diagnosis of carcinoma is made the typical cytological characteristics must be proven in the entire group of cells, whereas in all forms of prostatitis the atypias within cell aggregates occur almost exclusively locally.

In doubtful cases, intensive anti-inflammatory treatment with subsequent control biopsy is indicated. If it is still impossible to rule out carcinoma, punch biopsy, which permits evaluation of the stroma, is imperative as an aid to diagnosis.

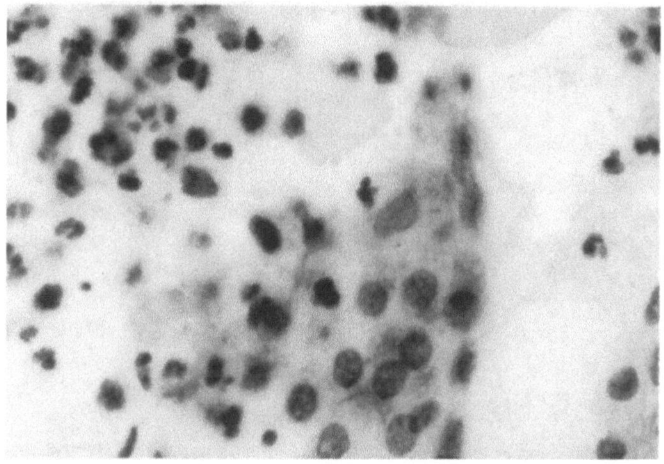

Fig. 174. Purulent prostatitis: partially dissociated sheet of prostatic epithelium with low-grade atypias (Pap II), surrounded by numerous leukocytes. ×400

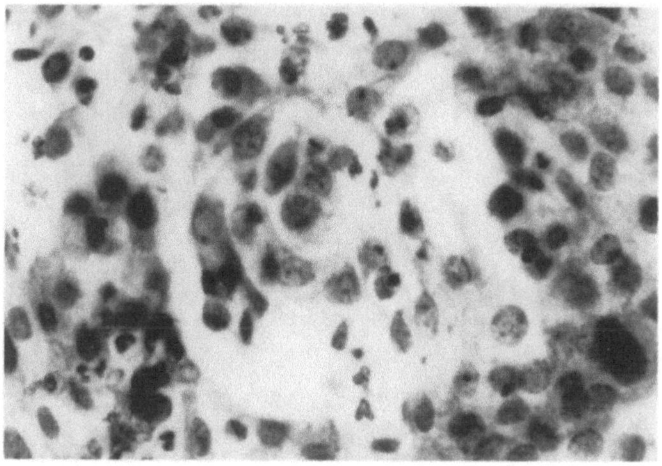

Fig. 175. Purulent prostatitis: large sheet of epithelial cells, infiltrated by leukocytes. ×400

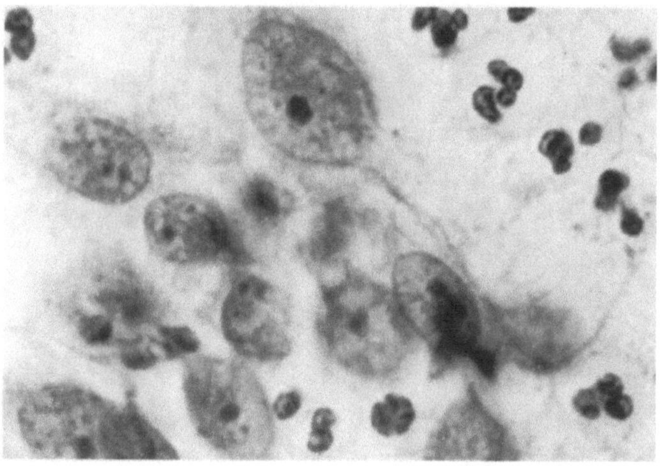

Fig. 176. Epithelial atypias (Pap III) in purulent prostatitis. Single nuclei display prominent nucleoli and disturbance of the nuclear arrangement. In some areas the chromatin structure is condensed. Nuclear membranes still intact. Oil immersion, ×1000

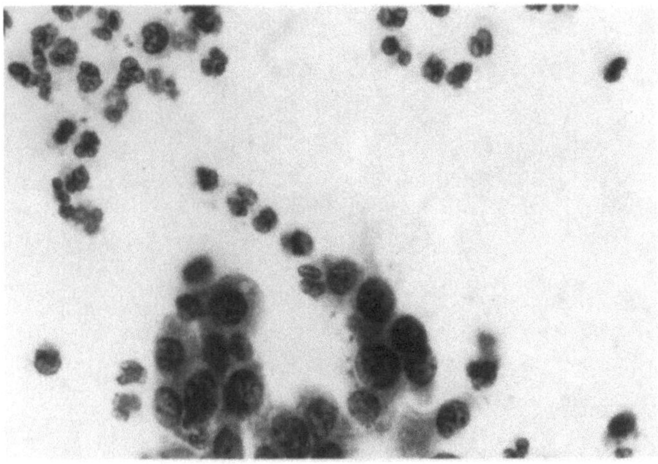

Fig. 177. Prostatic epithelium with atypias (Pap III) in purulent prostatitis. ×400

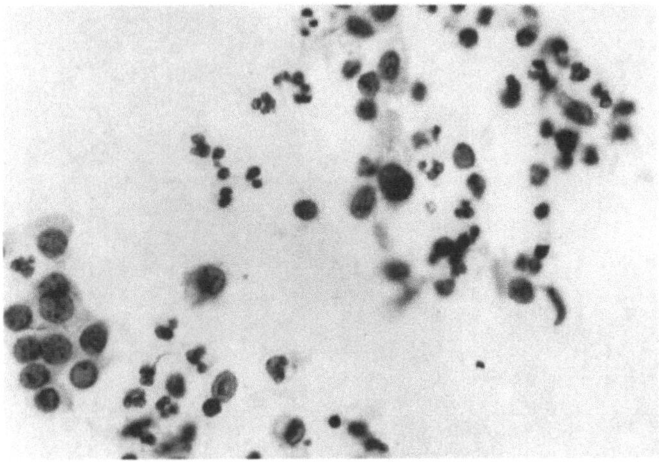

Fig. 178. Low-grade purulent prostatitis. ×400

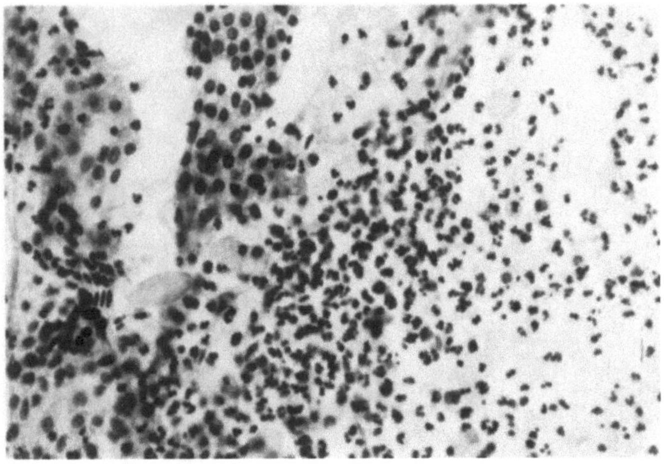

Fig. 179. Severe purulent prostatitis: the two sheets of prostatic epithelial cells (*left*) are surrounded and also infiltrated by numerous leukocytes. ×100

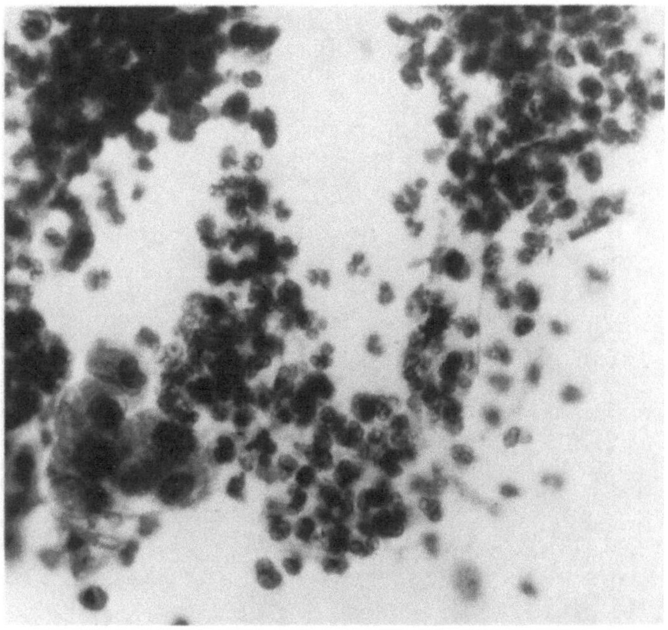

Fig. 180. Abscessing prostatitis with several clusters of histiocytes. ×400

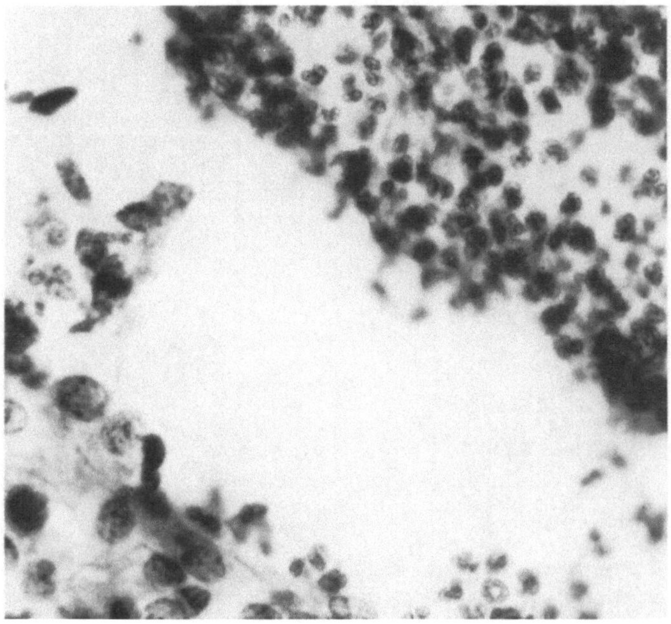

Fig. 181. Large cluster of histiocytes in abscessing prostatitis. ×400

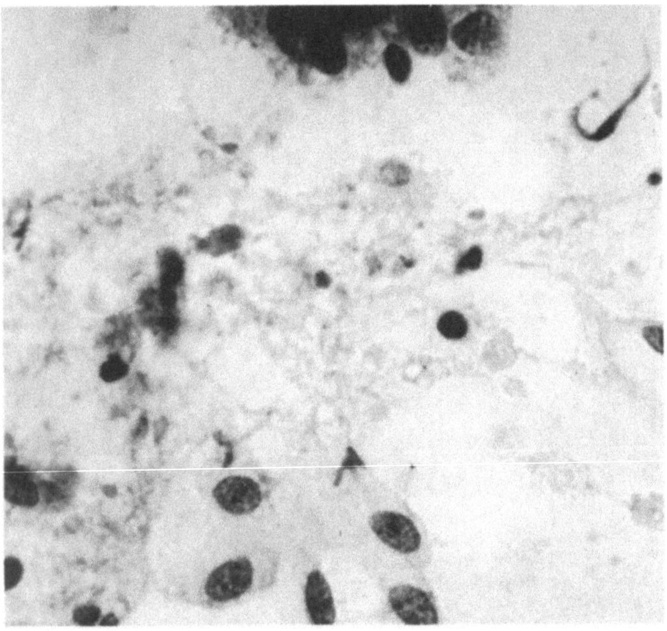

Fig. 182a. Mild chronic prostatitis: loose collection of round cells and histiocytes. A small part of a sheet of prostatic epithelium is seen at the *upper margin of the picture*. ×400

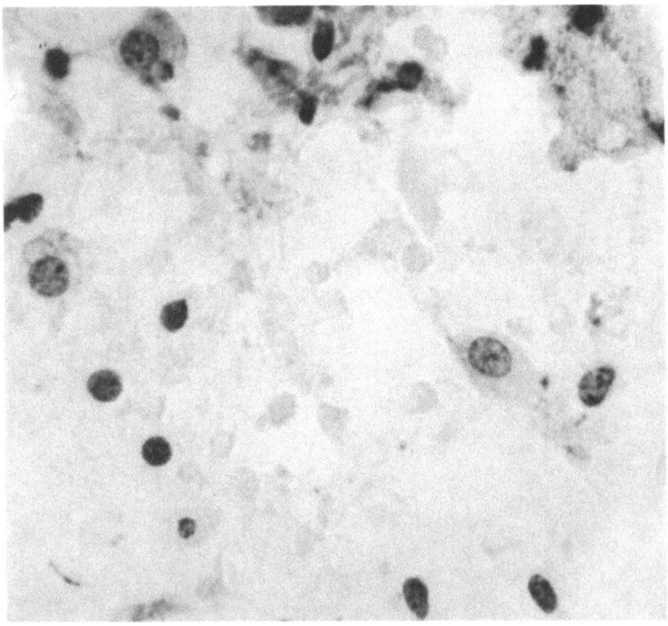

Fig. 182b. Same case: the histiocytes display a foamy, granular cytoplasmic structure as an expression of phagocytosis. ×400

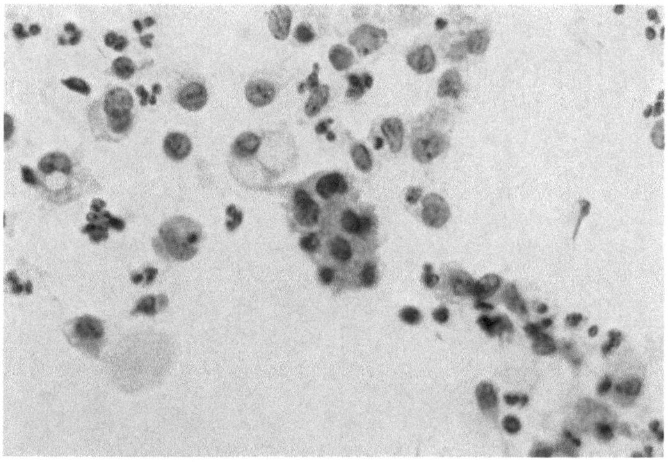

Fig. 183. Chronic recurrent prostatitis: small sheet of prostatic epithelial cells (*middle*), surrounded predominantly by histiocytes and by a few leukocytes and round cells. ×400

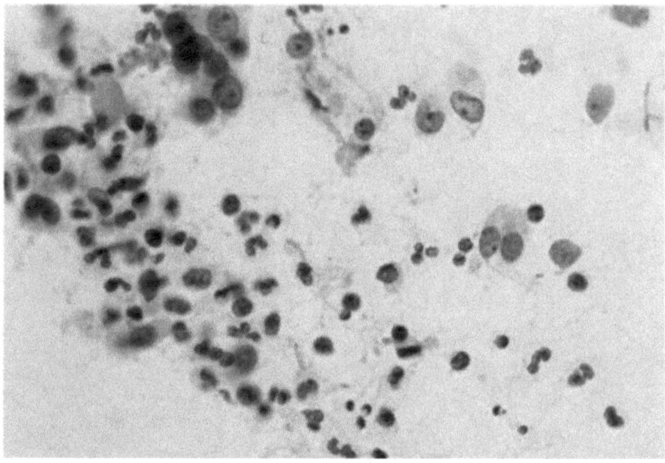

Fig. 184a. Chronic recurrent prostatitis with the typical variegated inflammatory cell pattern. ×400

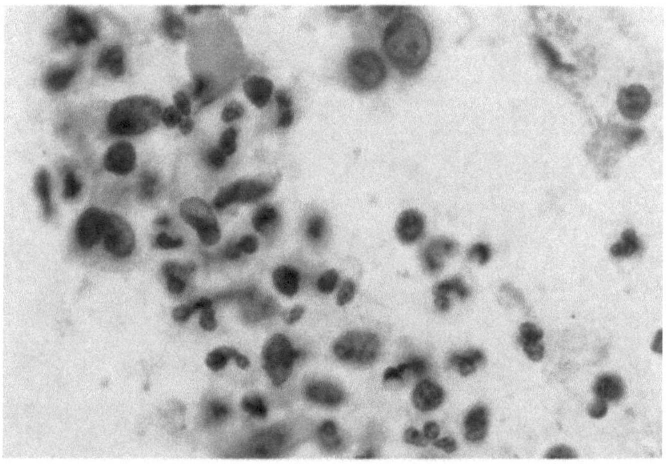

Fig. 184b. Same case at higher magnification. ×630

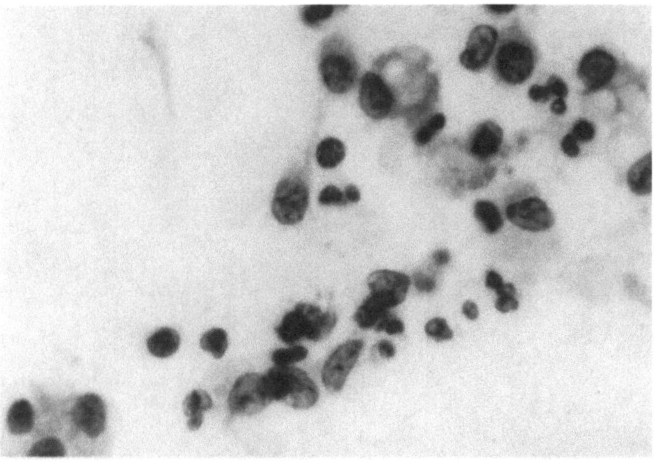

Fig. 185. Chronic recurrent prostatitis with a totally dissociated sheet of prostatic epithelial cells. ×400

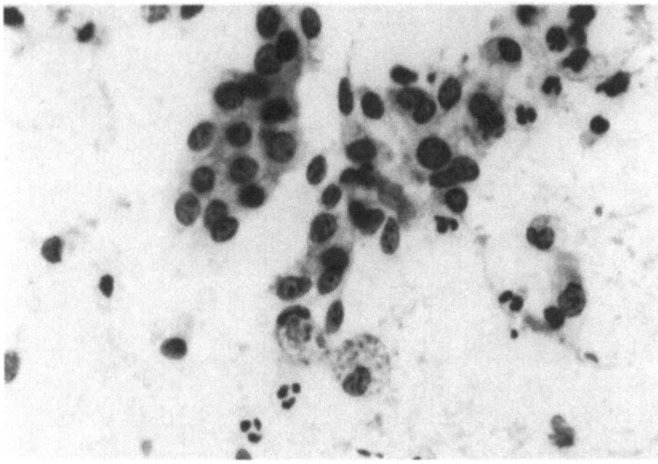

Fig. 186. Mild chronic recurrent prostatitis with only a low inflammatory cell content. ×400

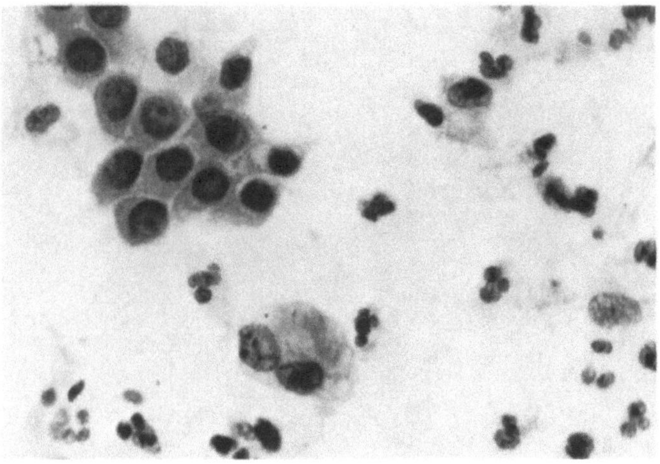

Fig. 187. Mild chronic recurrent prostatitis. ×400

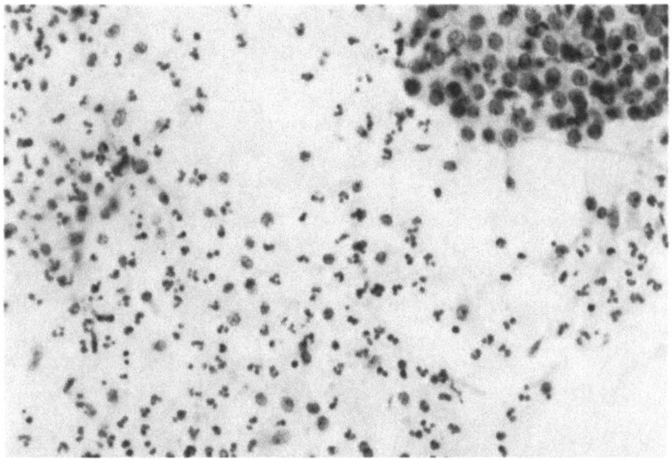

Fig. 188. Severe chronic recurrent prostatitis at the low screening magnification, with the typical variegated inflammatory cell pattern. ×100

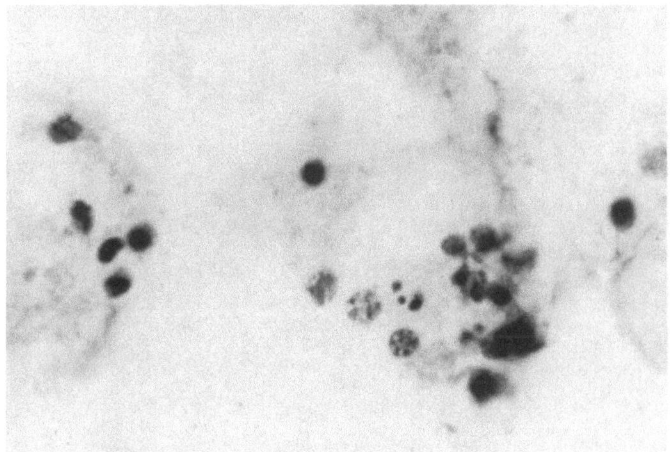

Fig. 189. Plasma cell with the characteristic spoked-wheel structure, in chronic recurrent prostatitis. ×400

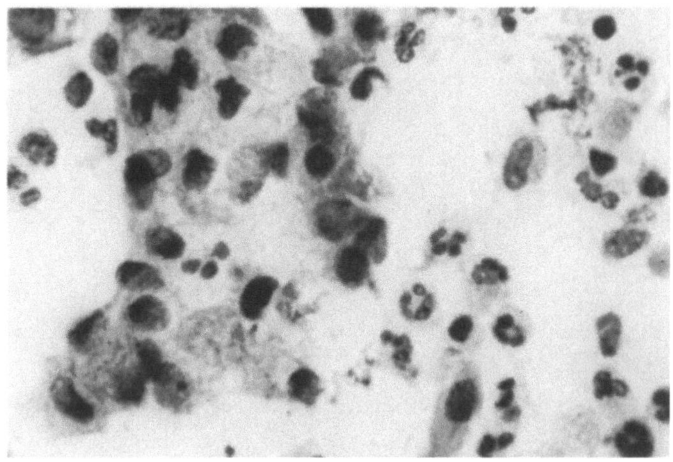

Fig. 190. Chronic recurrent prostatitis with marked dissociation of a sheet of prostatic epithelial cells. ×630

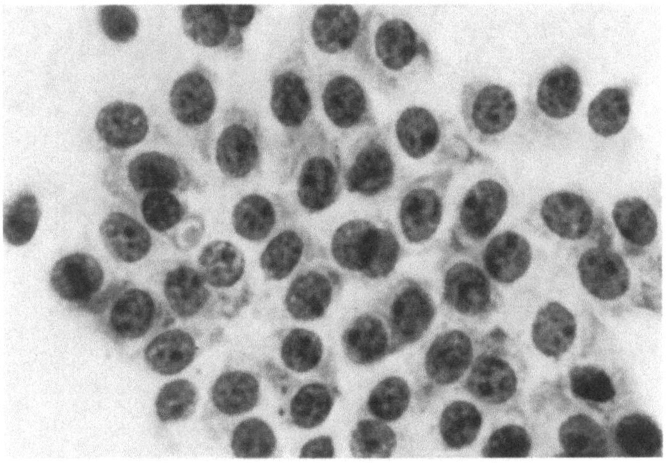

Fig. 191. Epithelial atypias (Pap III) in chronic recurrent prostatitis: many, though not all, of the nuclei display prominent nucleoli, some of which have lost their circular contour. × 630

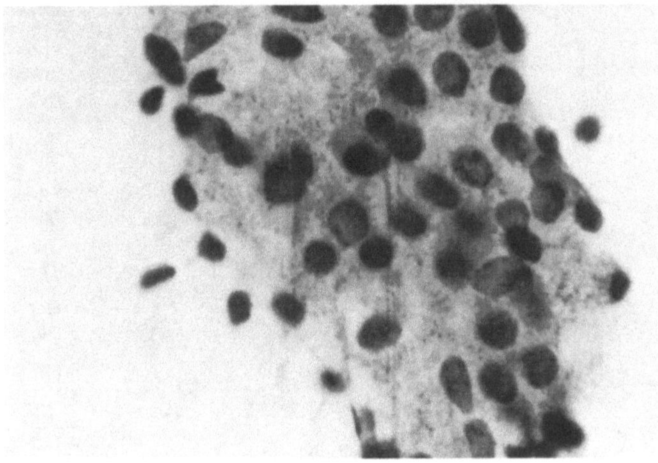

Fig. 192. Epithelial atypias (Pap III) in chronic recurrent prostatitis, with striking disturbance of the nuclear arrangement. × 630

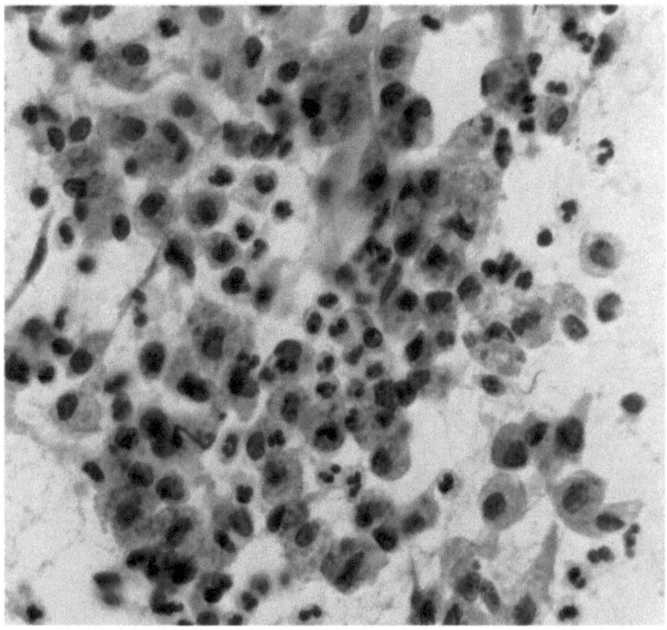

Fig. 193. Cytological appearance in prostatic infarction: severe chronic recurrent prostatitis and metaplastic squamous cells. ×400

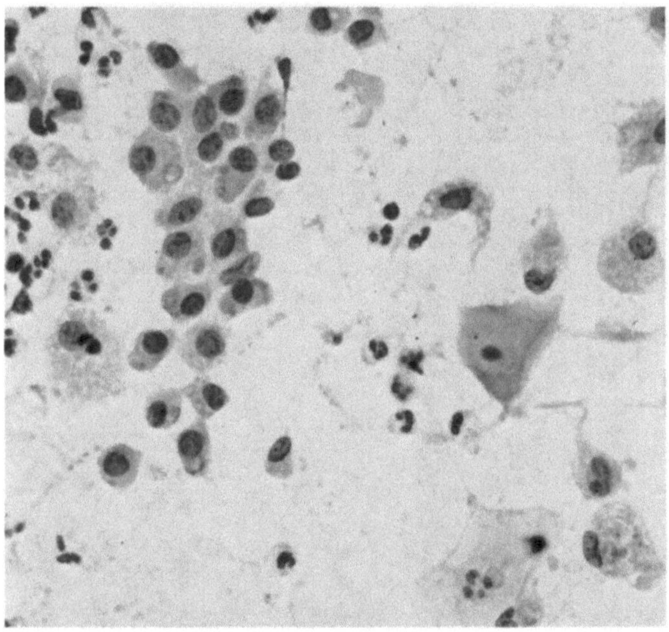

Fig. 194. Prostatic infarction with a low inflammatory cell content. ×400

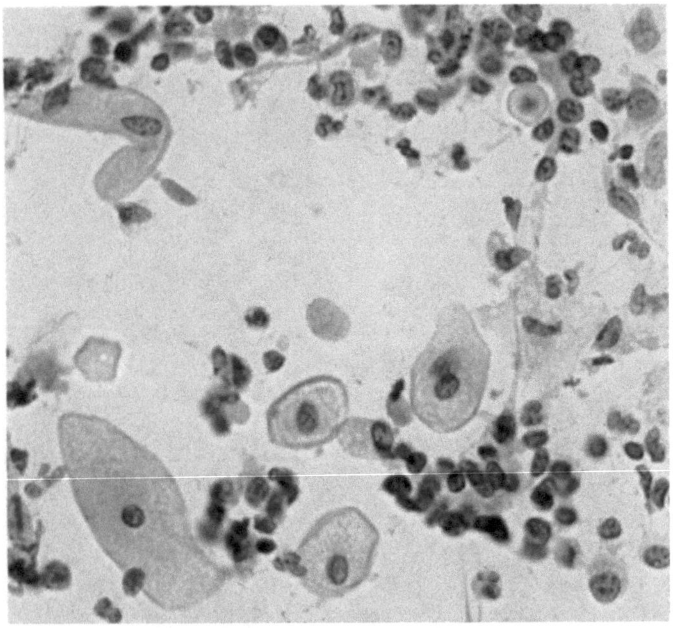

Fig. 195. Prostatic infarction, with numerous histiocytes and large squamous cells. ×400

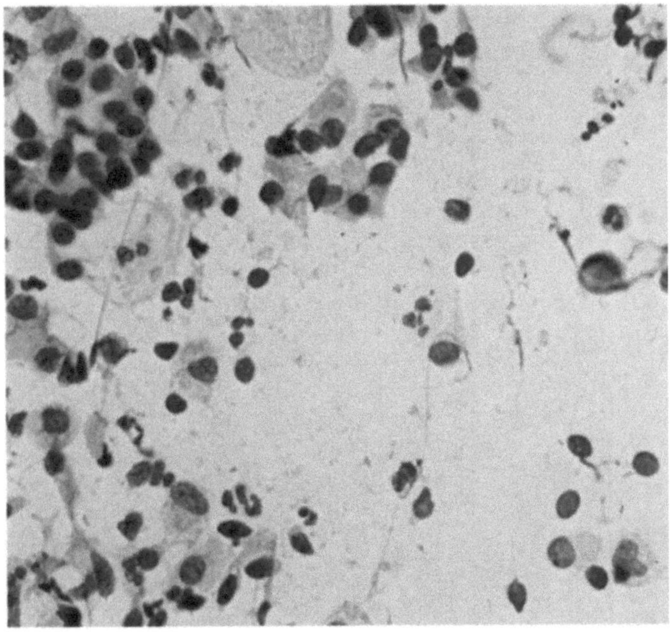

Fig. 196. Markedly dissociated, small sheet of prostatic epithelial cells in prostatic infarction. ×400

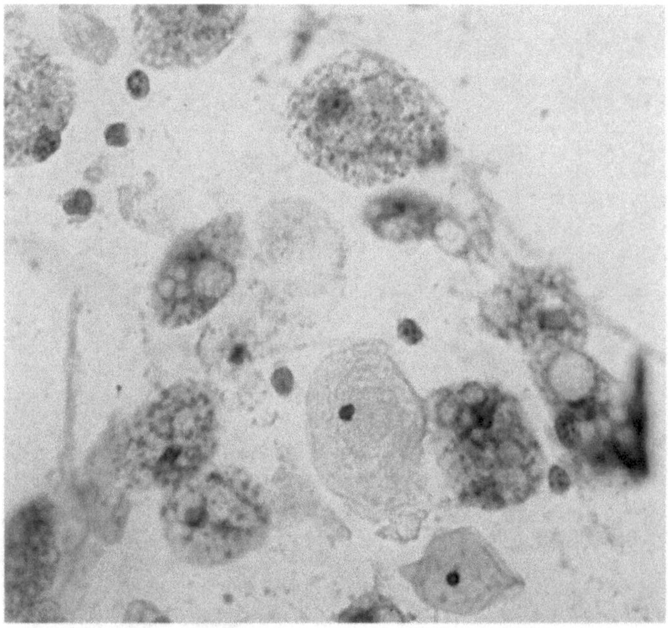

Fig. 197a. Numerous large foam cells and squamous cells in prostatic infarction. ×400

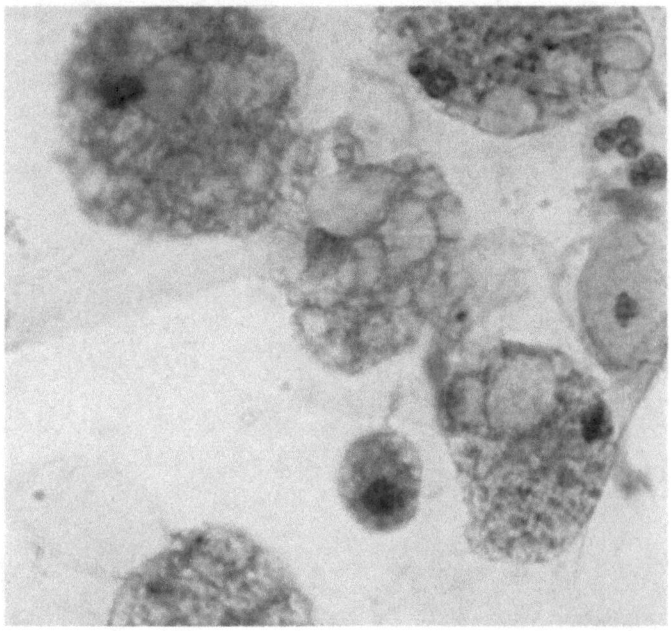

Fig. 197b. Same case at higher magnification. ×630

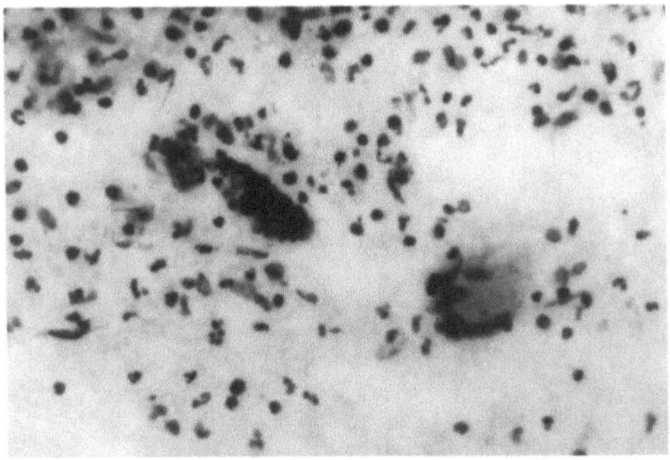

Fig. 198. Granulomatous prostatitis with Langhans' giant cells at the low magnification used for preliminary screening. ×100

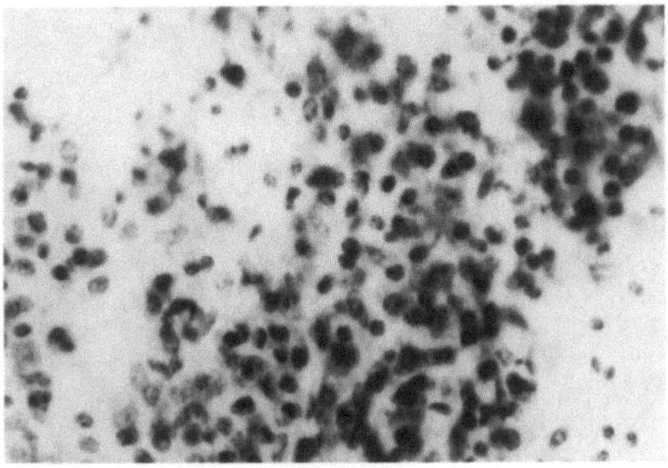

Fig. 199. Massive collection of inflammatory cells in granulomatous prostatitis. ×400

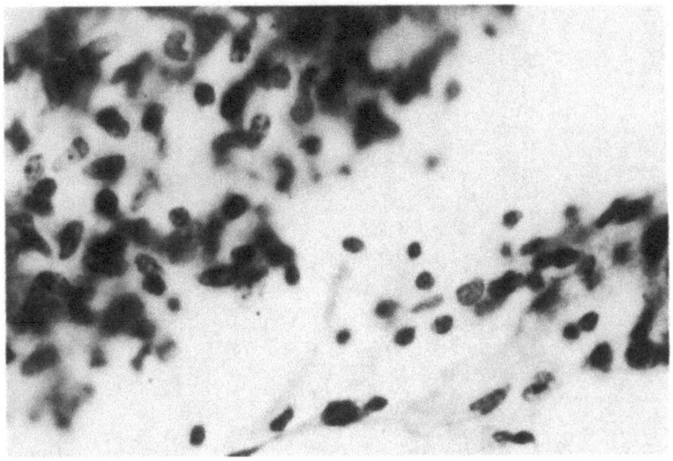

Fig. 200. Granulomatous prostatitis with a loose collection of inflammatory cells and clusters of histiocytes (*left of picture*). ×400

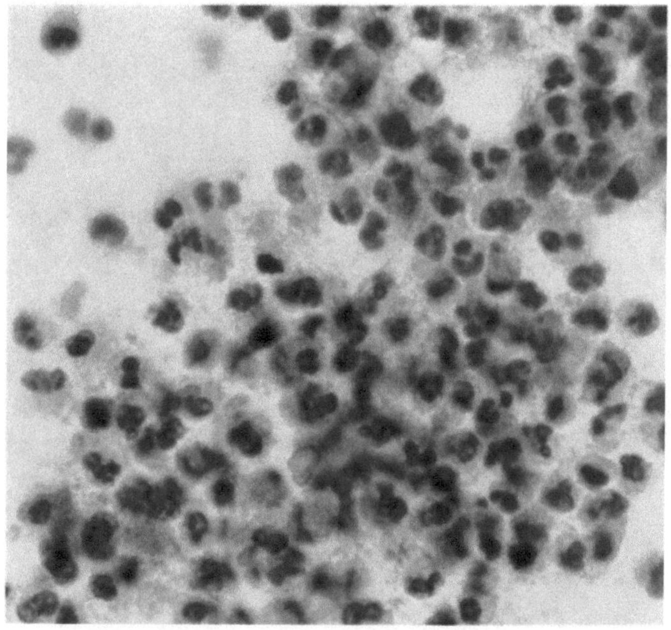

Fig. 201. Granulomatous prostatitis with a massive collection of leukocytes and histiocytes. ×630

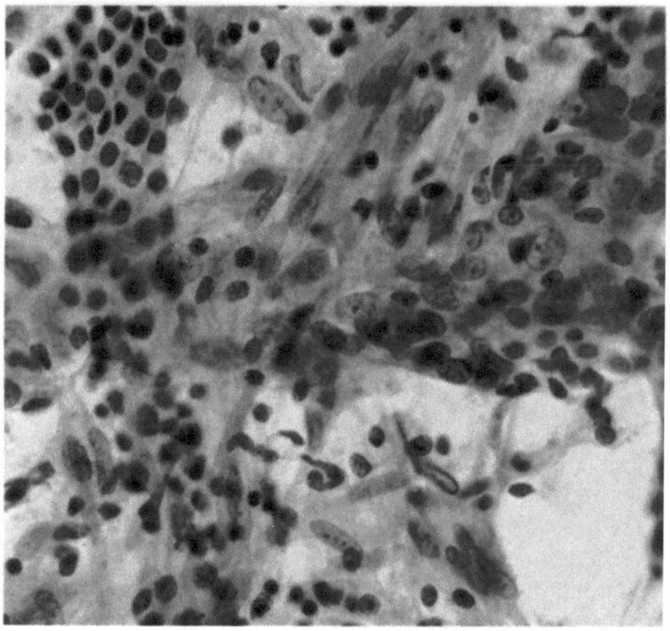

Fig. 202. Granulomatous prostatitis: prostatic epithelium (*upper left*), surrounded by clusters of histiocytes with transitional forms into epithelioid cells, particularly in the lower half of the picture. ×400

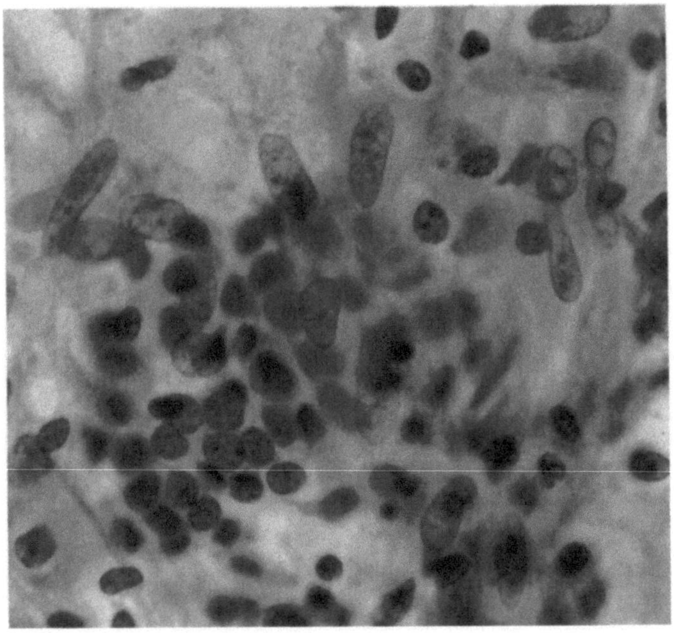

Fig. 203. Granulomatous prostatitis with clusters of histiocytes and localized epithelioid cells alongside a small, partially dissociated sheet of prostatic epithelial cells at higher magnification. ×630

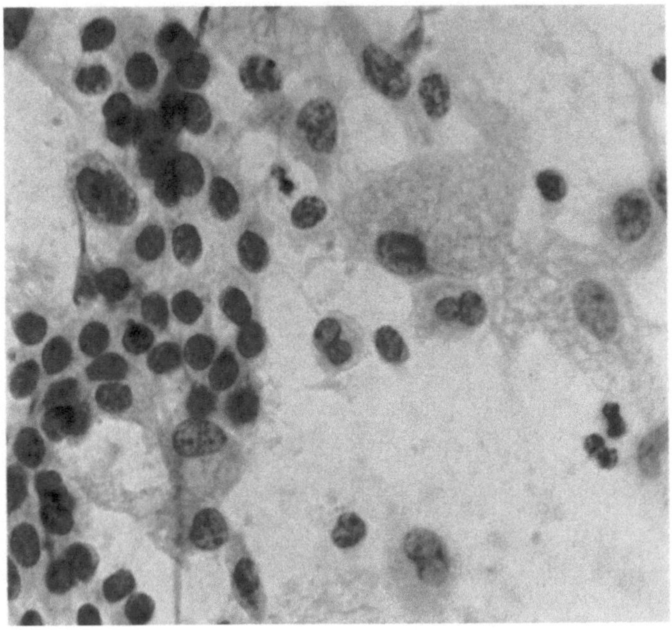

Fig. 204. Granulomatous prostatitis: sheet of prostatic epithelium, surrounded by histiocytes with abundant, foamy granular cytoplasm resulting from phagocytosis. ×630

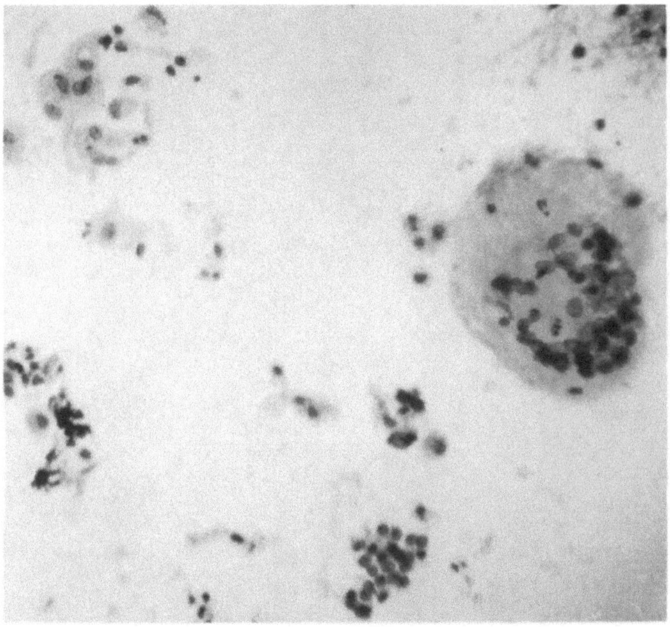

Fig. 205. Multinucleated histiocytic giant cell in granulomatous prostatitis, at the low magnification used for preliminary screening. × 100

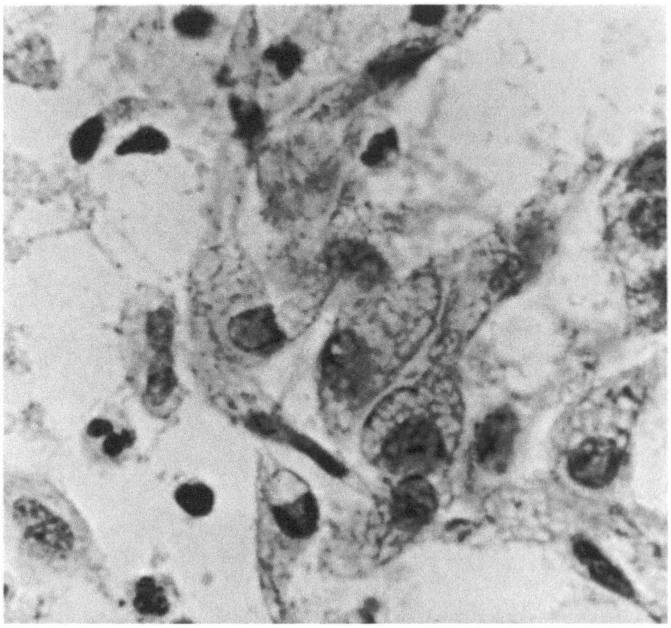

Fig. 206. Cluster of histiocytes with pronounced cytoplasmic vacuolation in granulomatous prostatitis. Oil immersion, × 540

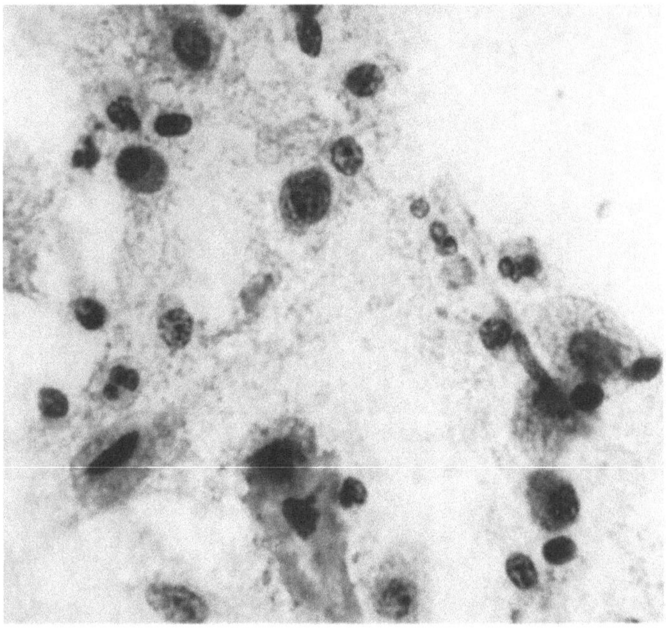

Fig. 207. Loose, mixed accumulation of inflammatory cells in moderately severe granulomatous prostatitis. ×400

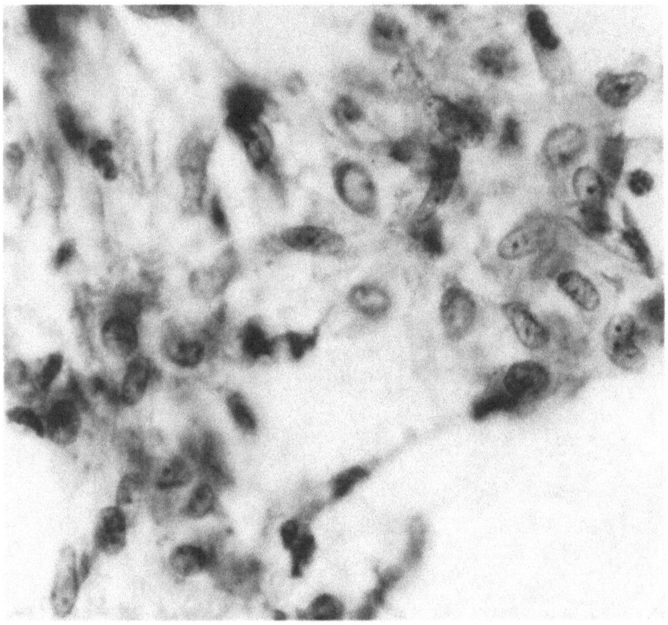

Fig. 208. Cluster of histiocytes with isolated transitions into epithelioid cells (*center*) in granulomatous prostatitis. ×400

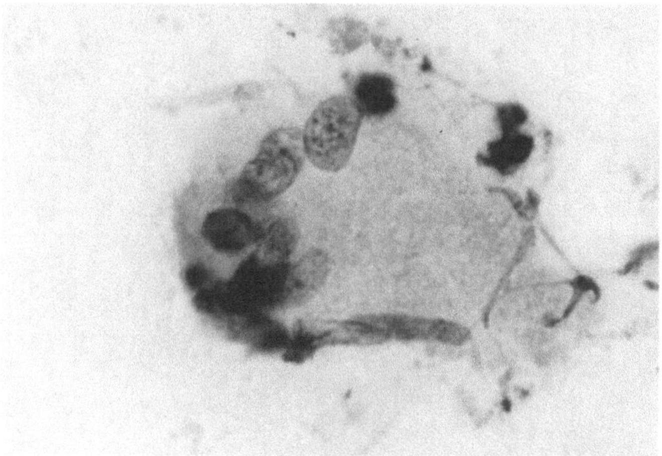

Fig. 209. Langhans' giant cells in granulomatous prostatitis. Oil immersion, × 540

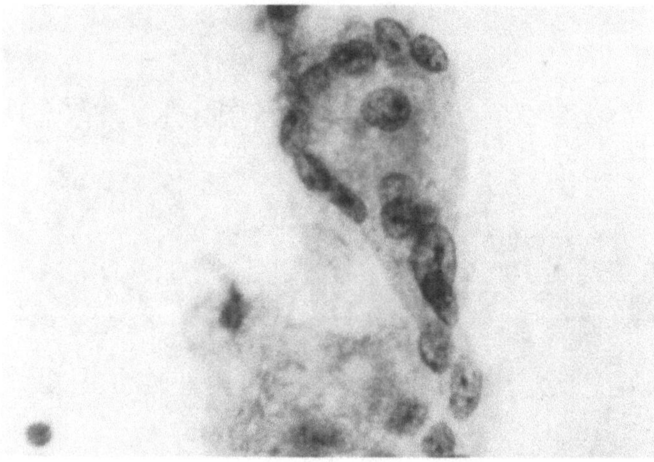

Fig. 210. Partially deformed Langhans' giant cell in granulomatous prostatitis. Oil immersion, × 540

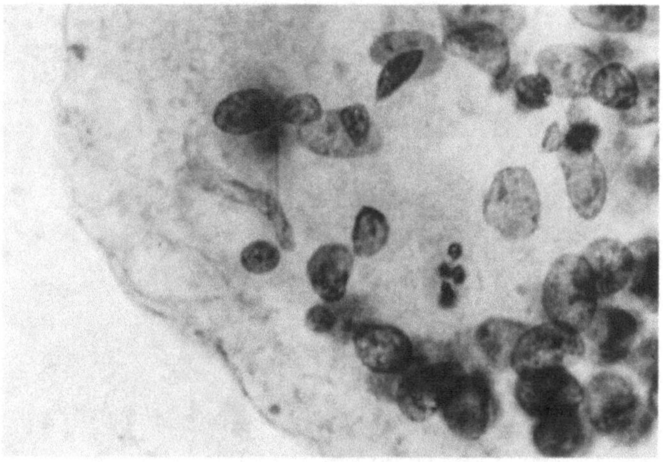

Fig. 211. Multinucleated histiocyte in granulomatous prostatitis (cf. Fig. 199). Oil immersion, × 540

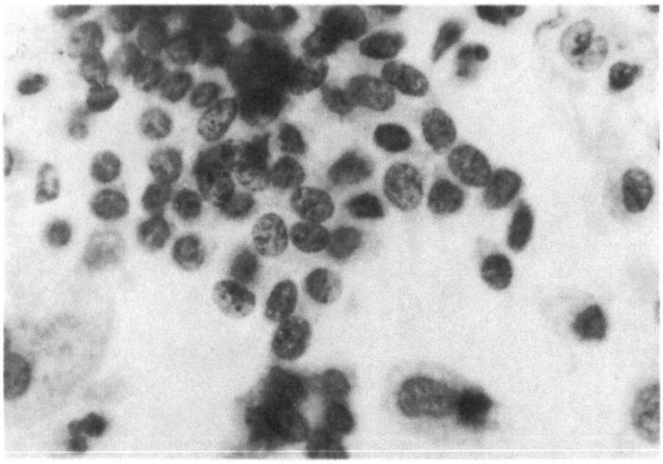

Fig. 212. Epithelial atypias (Pap III) in a partially dissociated sheet of prostatic epithelial cells in granulomatous prostatitis. ×400

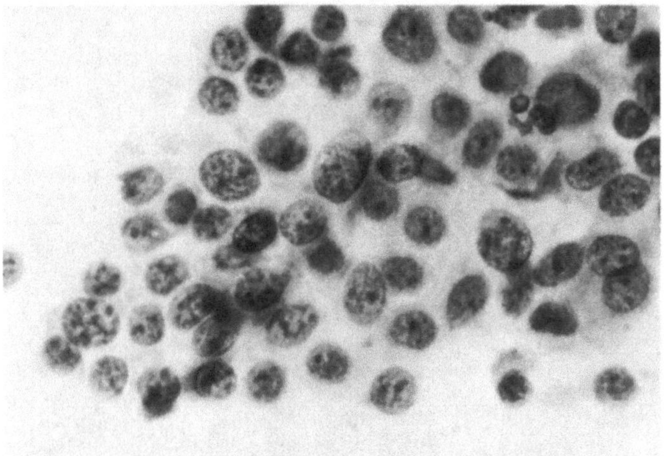

Fig. 213. Epithelial atypias (Pap III) in granulomatous prostatitis at higher magnification: moderate nuclear polymorphism and mainly conspicuous nucleoli. The chromatin structure is mostly loosely granular, nuclear membranes are easily recognizable. ×630

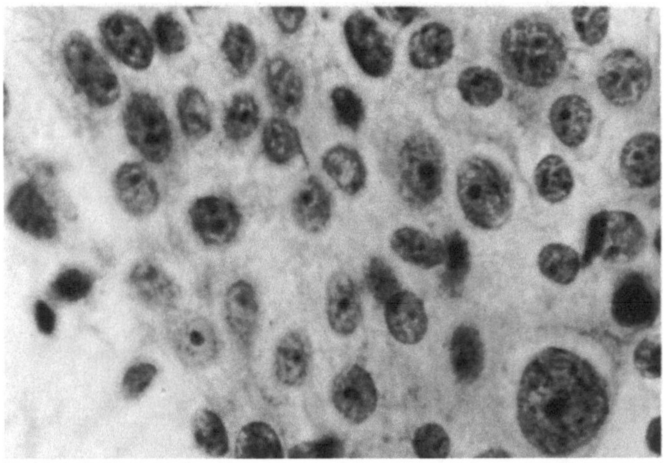

Fig. 214. Epithelial atypias (Pap IV) in granulomatous prostatitis: marked nuclear polymorphism with numerous prominent nucleoli, in some cases with loss of circularity; locally, more than one nucleolus per nucleus. Some of the nuclear membranes are no longer intact. ×630

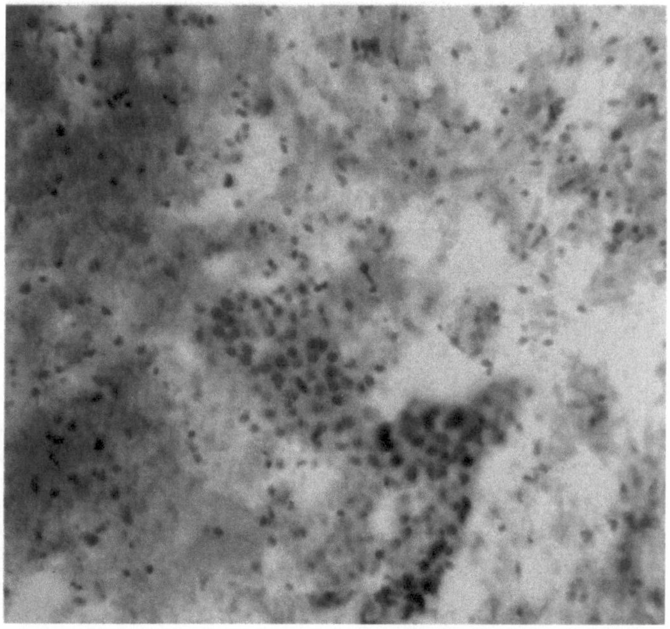

Fig. 215. Tuberculous prostatitis with the typical dirty background at the low magnification used for preliminary screening. ×100

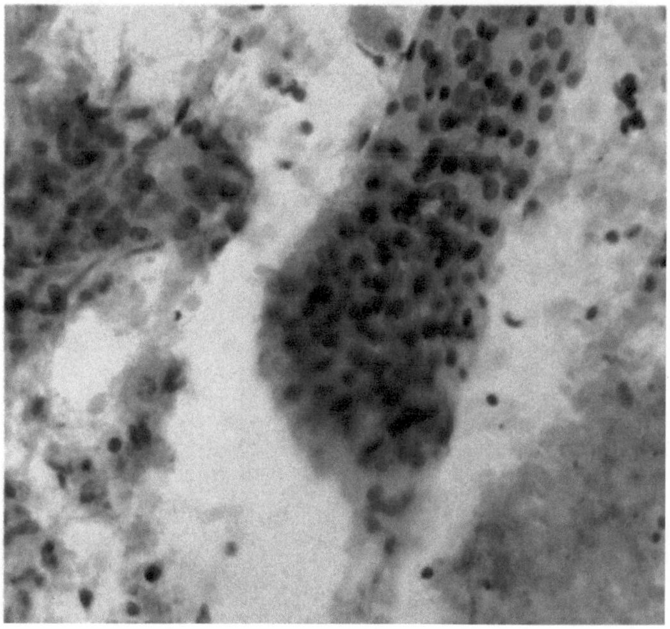

Fig. 216. Tuberculous prostatitis with a sheet of prostatic epithelium (*center*), clusters of histiocytes (*left*) and necrotic material (*right*). ×400

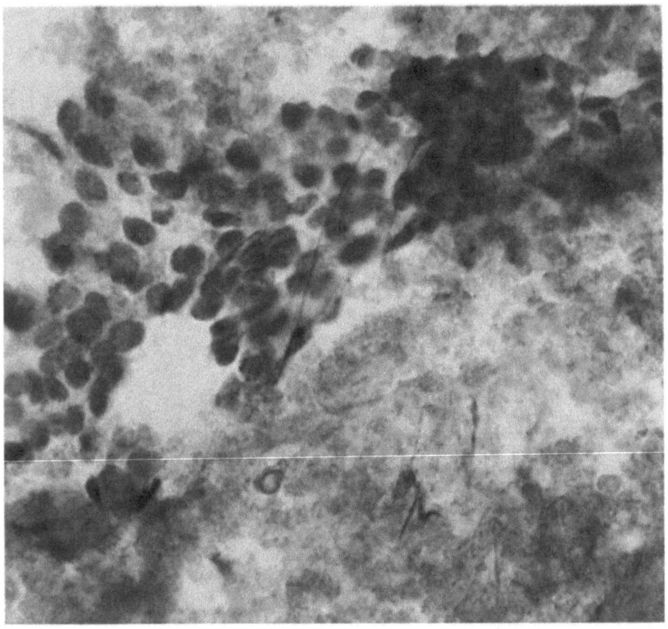

Fig. 217. Tuberculous prostatitis with a partially dissociated sheet of prostatic epithelial cells and abundant necrotic material. ×400

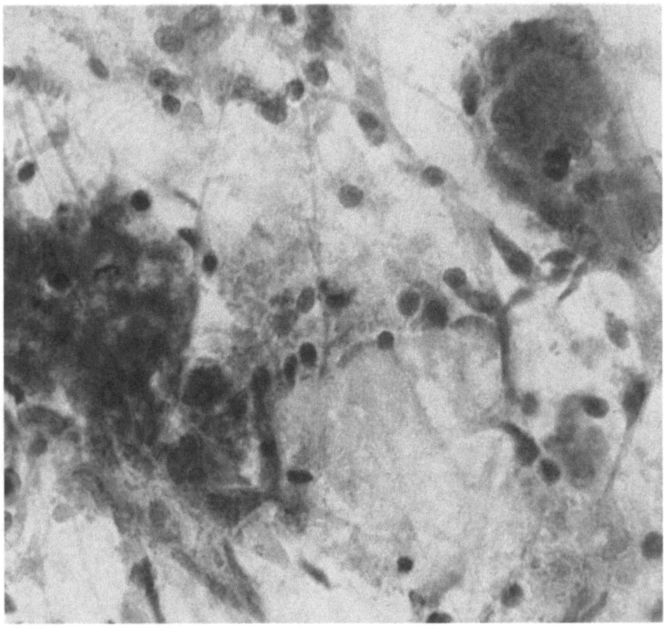

Fig. 218. Tuberculous prostatitis with two clusters of histiocytes within necrotic material. ×400

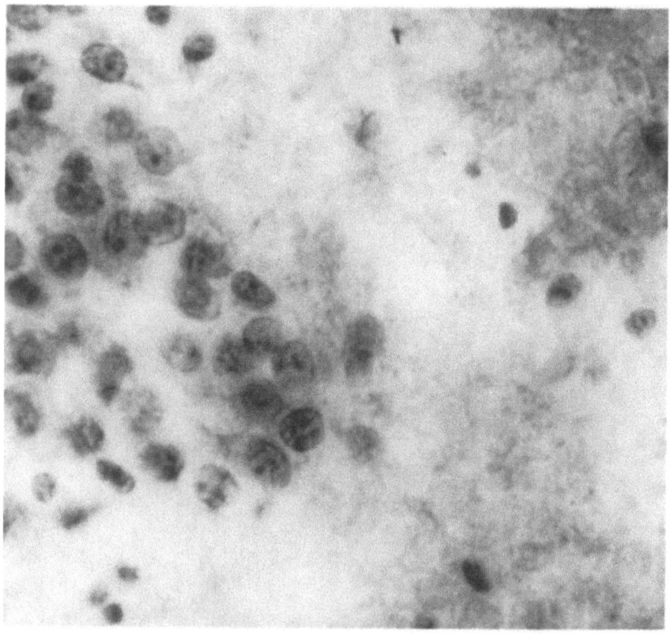

Fig. 219. Tuberculous prostatitis with a markedly dissociated sheet of prostatic epithelial cells, atypias (Pap III) and necrotic material. ×400

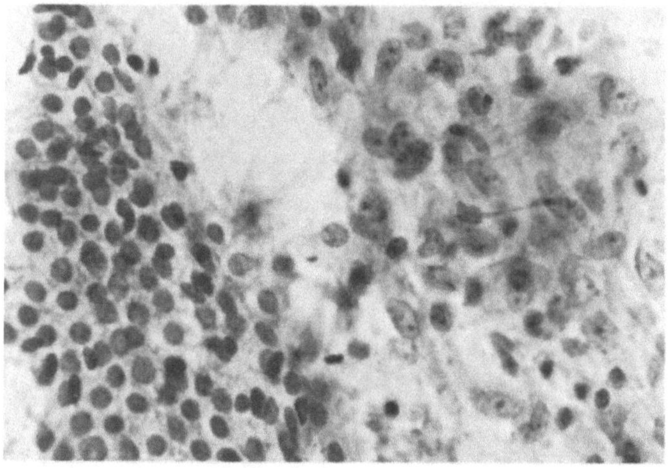

Fig. 220. Tuberculous prostatitis with prostatic epithelium (*left*) and a large cluster of histiocytes with transitions into epithelioid cells (*right*). ×400

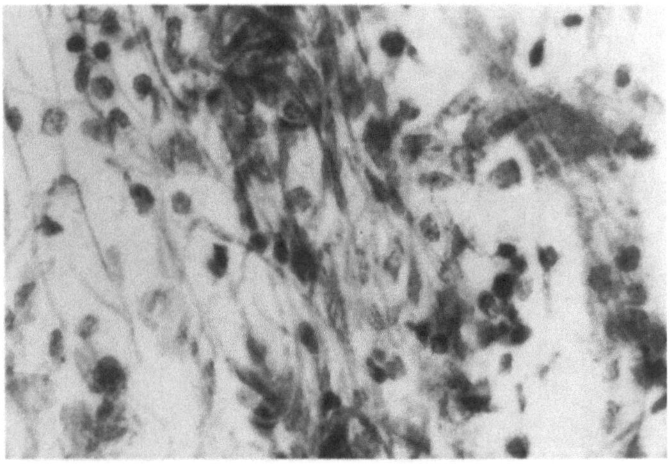

Fig. 221. Tuberculous prostatitis with a variegated inflammatory cell pattern and a cluster of histiocytes (*upper half of picture*). ×400

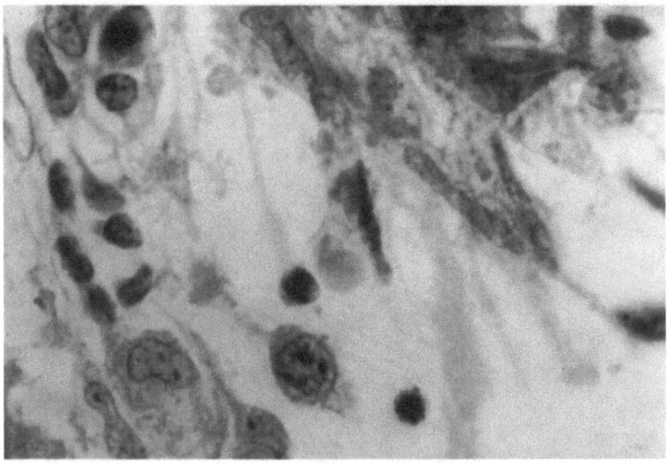

Fig. 222. Tuberculous prostatitis with a cluster of epithelioid cells (*upper right*). ×630

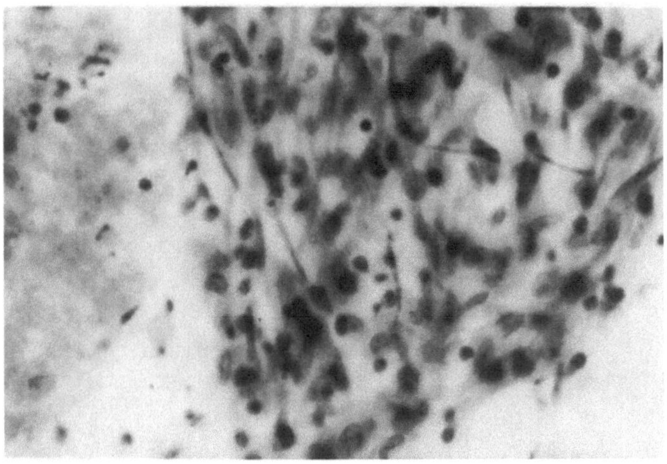

Fig. 223. Tuberculous prostatitis with a large cluster of epithelioid cells alongside necrotic material. ×400

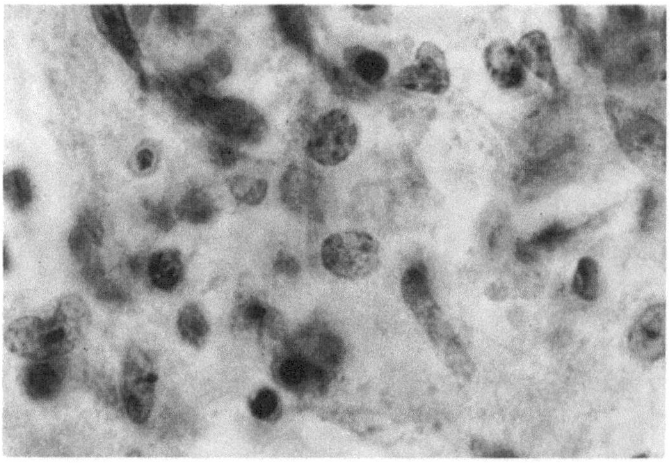

Fig. 224. Tuberculous prostatitis with a variegated inflammatory cell pattern in necrotic material. Oil immersion, × 540

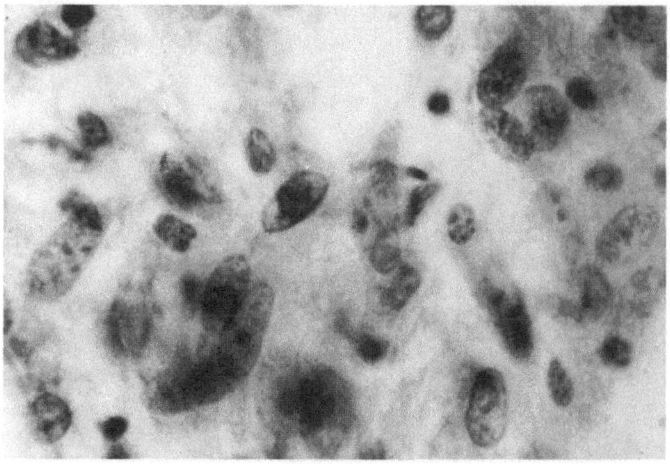

Fig. 225. Tuberculous prostatitis with epithelioid cells (*left*) adjacent to clusters of histiocytes (*right*). Oil immersion, × 540

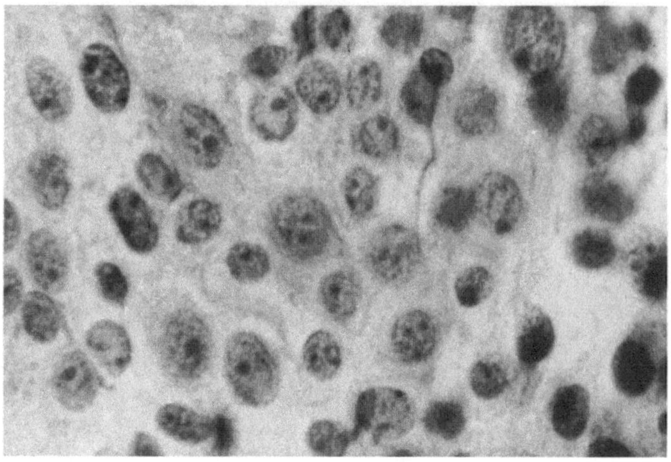

Fig. 226. Epithelial atypias (Pap IV) in tuberculous prostatitis (cf. Fig. 208). × 630

13 DNA Cytophotometry

Whereas *cytomorphological examination* of cells and nuclei permits *qualitative* assessment of prostatic disease only, *cytophotometry* allows *quantitative* assessment of the biological behavior, i.e. the malignancy potential, of the prostatic carcinoma before and during treatment with all known forms of therapy, by means of detailed analysis of the nuclear DNA. *Cell kinetics* can also only be determined by quantitative measurement of the *nuclear DNA content,* i.e. by cytophotometry and not by *autoradiography* or *liquid scintillation spectrophotometry,* which are methods of measuring DNA synthesis by incorporating radioactive-labelled thymidine (H^3).

In cytophotometry the individual cell nuclei are assigned to the appropriate phase of the cell cycle according to their DNA content. Thus with the aid of the DNA cytophotogram the frequency distribution of the cycle phases in a measured cell population can be established. Changes in cell kinetics accordingly result in changes in the cytophotograms obtained by determining the level of DNA, since, as a result of the change in the number of chromosomes, tumor cells exhibit a higher DNA content than cells in normal tissue, i.e. aneuploidy (SANDRITTER, 1952; LEUCHTENBERGER et al. 1954).

This can be seen in numerous organs, e.g. in the urogenital tract in the *bladder* (LEWI et al. 1969; LEDERER et al. 1972) and in the *prostate gland,* particularly in *carcinoma of the prostate* (SPRENGER et al. 1974, 1976); ZETTERBERG and ESPOSTI 1976; BICHEL et al. 1976; LEISTENSCHNEIDER and NAGEL 1979, 1980, 1982, 1983; BÖCKING 1981; SEPPELT and SPRENGER 1981; ESPOSTI 1982).

In cytophotometry the cell or organelle is looked upon as a cuvette filled with a fluid which is irradiated with monochromatic light and whose absorption is then measured.

The first cytophotometric absorption measurements were conducted by CASPERSSON (1936), who measured the absorption of ultraviolet light by the structures of the cell nucleus.

FEULGEN and ROSSENBECK (1924), developed a method known as *Feulgen's reaction* for staining nuclear chromatin, which is not only specific for DNA but also quantitatively accurate.

All cytophotometric determinations of nuclear DNA are based on Feulgen's reaction. They can be divided into 2 method systems:

- *single-cell scanning cytophotometry* (fluorometry, absorption photometry)
- *flow-through cytophotometry.*

13.1 Feulgen's Reaction

The reaction is dependent on the fact that fuchsin treated with sulphur dioxide (Schiff's reagent) reacts with 2 aldehyde groups, which must be at least 10 Å apart, to form a scarlet-colored compound. The DNA in the nucleus has a scalariform structure; the basic structure is formed by desoxyribose and phosphoric acid. The purin base can be removed from this structure by means of acid hydroly-

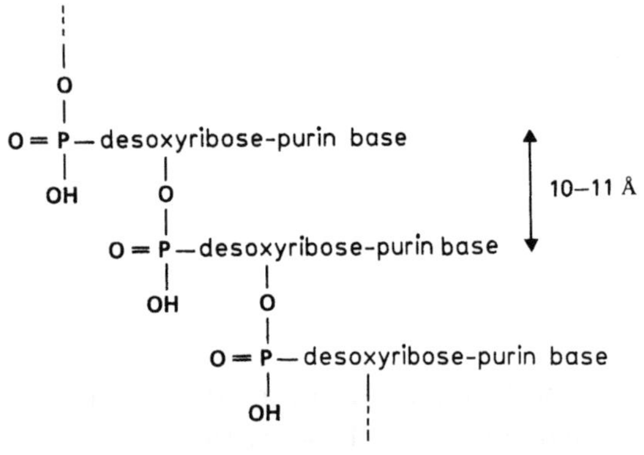

Fig. 227. The structure of DNA (BURCK 1966)

sis, free aldehyde groups being formed at intervals of 10.2 Å **(Fig. 227).** The Schiff's reagent is deposited on the "purin-free staircase" and reacts with the aldehyde groups to produce a scarlet coloration (BURCK 1966).

13.2 Single-cell Scanning Cytophotometry

13.2.1 Fluorometry

The object to be measured by fluorometry is rendered fluorescent by irradiation with ultraviolet, violet-blue or green light. It thereafter emits fluorescent light of equal intensity in all directions **(Fig. 228),** a corresponding portion of which is absorbed by the microscope objective. If it is the quantity of a material which is to be measured, e.g. the DNA content of the nucleus, the intensity of the fluorescent light emitted and collected from all points of the object is then measured, preferably at the exit pupil of the microscope (known as the pupil plane). A diaphragm in the image plane of the object reduces the measuring instrument to the size of the object to be studied.

If reliable results are to be obtained the fluorescing materials (fluorochromes) must be tested for linearity at the concentrations and layer thicknesses known to occur (RUCH 1979).

The best results are obtained if the DNA to be measured is stained with the following fluorochromes:

BAO
acriflavine-Schiff's reagent
auramine O

The fluorescence of the cells or nuclei to be measured always fades during preparation. However, this can be prevented by using phase contrast microscopy. In order to keep this fading to an absolute minimum it is important that irradiation with UV light be as short as possible. The cytofluorometers con-

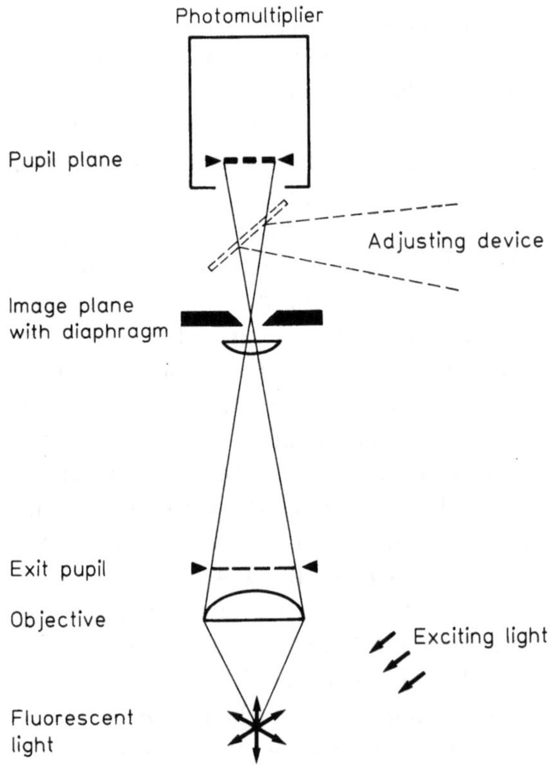

Fig. 228. The principle of cytofluorometry (RUCH 1979)

ventionally employed at present fulfil this requirement as they have electrically controlled shutters which guarantee excitation times ranging from a few milliseconds down to nanoseconds.

The values thus measured are *relative values* for the specific quantity of material measured and it is thus possible to compare the amounts of material in different objects. *Absolute values* are obtained by comparison with a standard population, e.g. liver cell nuclei, leucocytes or sperm.

Fluorometry can also be used to measure total proteins, arginine, lysine and biogenic amines.

Quantitative determination by fluorometry is now as good as absorption microscope photometry as a method of analyzing cells. In addition to DNA histograms, fading curves and line scans can also be produced with the aid of a process computer (REUTER 1979).

13.2.1.1 Equipment Required for Cytofluorometry

- Fluorescence microscope with stabilized light source
- Photometer
- Computer (process computer)

13.2.1.2 Stains

If optimal fluorescence is to be obtained *fluorochromes* meeting the following requirements should be used:

- high fluorescence intensity,
- high specificity,
- minimum fading on exposure to the exciting radiation,
- spectral characteristics as suitable as possible for the wavelengths of the lasers employed (WITTEKIND 1979).

The fluorescence of the fluorochromes must be higher than the primary fluorescence of the preparations, which should be treated as stained preparations but without adding the fluorochromes. The intensity of fluorescence must be proportional to the amount of bound fluorochrome.

High *specificity* is attained if the fluorochrome exhibits a high degree of affinity for a certain biopolymer and thus a distinct fluorescent color is produced which is clearly distinguishable from the reactions of other adjacent macromolecules.

13.2.1.3 Fluorescent Staining Technique

Fixation. The following solutions are suitable for fixation for fluorescence microscopy and cytofluorometry:

methanol
alcohol
alcohol-acetic acid mixtures
neutralized formalin
(partial blocking of reactive amino groups)

Glutaraldehyde is not recommended since its free aldehyde groups are too small. Fixatives for fluorometry containing heavy metals are also unsuitable (RUCH and LEHMANN 1973).

The *fixation times* for the different substrates in which DNA is to be measured vary as follows:

cell suspensions 4–24 hours
cytological smears 15–30 hours
thin blocks of tissue
(2 mm) at least 24 hours

Following *fixation* carefully rinse off the fixative. Use distilled water after *fixation with formalin* and alcohol after *fixation with mixtures containing acetic acid.*

The *cellular material can be stored* at 4 °C suspended in 70% alcohol (RUCH and LEHMANN 1973). Storage for longer than 3 weeks is not to be recommended as both the DNA and the protein levels are reduced.

13.2.1.4 Staining Techniques

BAO Stain. If the fluorochrome BAO is used instead of pararosaniline a modified Feulgen's reaction of the DNA can be induced (RUCH 1970). This reaction can also be induced after fixation with methanol, alcohol-acetic acid or formalin. *The chromatin structure exhibits a brilliant blue fluorescence.*

Fixation	
Methanol (slides must be grease-free!)	15 min at room temperature *or* 30 min at 4 °C
Rinse (distilled water)	2 min
Hydrolysis	
1 N HCl	8 min, 60 °C *or*
6 N HCl	20 min at room temperature
Rinse (distilled water)	5 min
Staining	
(100 ml aqueous BAO solution, 0.01%; 10 ml 1 N HCl, 5 ml $NaHSO_3$, 10%) (Always prepare fresh solution)	2 h
Rinse (distilled water)	2 min
Sulfite solution (180 ml dist. water, 10 ml 1 N HCl, 10 ml $NaHSO_3$, 10%) (This solution will keep for 1 week)	3 × 2 min
Running water	10 min
Dehydration	30 s with 50%, 70%, 95%, 100% and 100% ethyl alcohol respectively
Xylene	2 × 30 s

Mount in Fluormount[1]

Excitation: UV (excitation filter: UG 1)

Result: DNA emits blue fluorescence

1 Manufactured by Nordwald, Hamburg and Fluka, Basel

Acriflavine-Schiff's reagent. A fluorescent Schiff's reagent is employed for this reaction, which is highly specific for DNA. It gives a more brilliant coloration against a dark background than the conventional Feulgen's fluorescence reaction.

Fixation

Carnoy's solution	15 min
Composition of the fluorescent dye	
Acriflavine hydrochloride	1 g
Potassium metabisulfite	2 g
Distilled water	200 ml
1 N HCl	20 ml
Dissolve acriflavine and potassium metabisulfite in distilled water and then add HCl. The solution is basically stable and should be left to stand overnight before use.	
Rinse (distilled water)	1 min

Hydrolysis

1 N HCl	15–30 min (20 °C) *or* 10 min (60 °C)
Rinse (distilled water)	1 min

Staining

Acriflavine-Schiff's reagent	20 min
Rinse with HCl-alcohol 1% HCl, 95% alcohol, to remove the non-reactive portion of the fluorochrome, instead of SO_2 water	5 min
fresh HCl-alcohol	10 min
dehydrated alcohol	3 × 1 min
xylene	1 min

Mount in Fluormount[1]

Excitation: UV (excitation filter: BG 12)

Result: DNA emits gold-yellow fluorescence

1 Manufactured by Nordwald, Hamburg and Fluka, Basel

Auramine O stain. If the object has an intrinsic blue fluorescence auramine O stain, which fluoresces a yellowish-green color, should be employed.

Fixation	15 min
Rinse (distilled water)	2 min
Hydrolysis (see BAO stain)	
Rinse	5 min

Staining

Stain with freshly prepared auramine O-SO_2 solution (10 ml auramine O solution, 0.2% in distilled water, 1 ml HCl, 0.5 ml 10% $NaHSO_3$ solution (fresh), 0.025 g activated charcoal, shake briefly and filter).	3 h
Rinse in water	a few seconds
Wash in sulphite water (see BAO stain)	3×2 min
Wash in running water	10 min
Dehydrate with the alcohol series (see BAO stain)	

Mount in Canada balsam or glycerin

Excitation: UV (excitation filter: UG 1)

Result: DNA fluoresces a brilliant yellowish-green. The preparations can be stored for 3–5 days in a refrigerator before measurement.

13.2.2 Absorption Scanning Cytophotometry

This type of cytophotometry is also conducted with single cells!

Whereas the fluorescence values measured by fluorometry, which are directly proportional to the quantity of material, can be used for statistical evaluation without further transformation, this is not possible in the case of absorption cytophotometry if distributions of material within an object, for example the DNA content of a cell nucleus, are to be quantified. Thus for absorption cytophotometry a scanning procedure is required which permits the object (a single nucleus) to be systematically transported past a scanning element by means of a mechanically operated fine scanning stage. The scanning element is basically nothing more than a photomultiplier with a diaphragm in front of it. The object is moved under the objective with the aid of the fine scanning stage in defined stages, in either a meandering or a comblike movement and thus measured by the cytophotometer.

The *scanning stage is controlled* by a process computer which gives the signals for measurement. At the same time the computer controls the

- scanning stage,
- the monochromator of the photometer and
- the shutters with which the degree of illumination in the microscope is regulated

according to the program via an interface.

The *process computer* also stores the extinction values measured and prints out the corresponding histogram or cytophotogram on a downstream printer. The basic statistical data are printed out simultaneously:

mean value,
standard deviation and
coefficient of variation.

Figure 229 shows the equipment required for microscope scanning photometry as set up in the laboratory.

Single-cell scanning photometry is *time-consuming* since it takes at least 2 hours to measure the DNA in about 100 nuclei. Moreover, the technician must identify each cell as carcinomatous and place it in the measuring instrument *before* measurement!

The time disadvantage is outweighed by the advantage afforded by the fact that it is possible to carry out qualitative cell morphology and quantitative measurement of nuclear DNA in the same material and to match the results.

Furthermore, since the DNA is measured in each individual cell it is not possible to falsely identify the DNA pattern in the aspirate since only cells which are unequivocally carcinomatous are marked before measurement, whilst the benign adenoma cells, inflammatory cells and fibroblasts are recognized as such, are not marked and are thus not included in the measurement.

The measurement of DNA by single-cell scanning photometry or fluorometry is eminently suitable for monitoring conservative treatment of carcinoma of the prostate, since it guarantees measurement of carcinoma cells alone. The results obtained can be reproduced at any time since the position of the cells on the slide remains unaltered because they have been fixed.

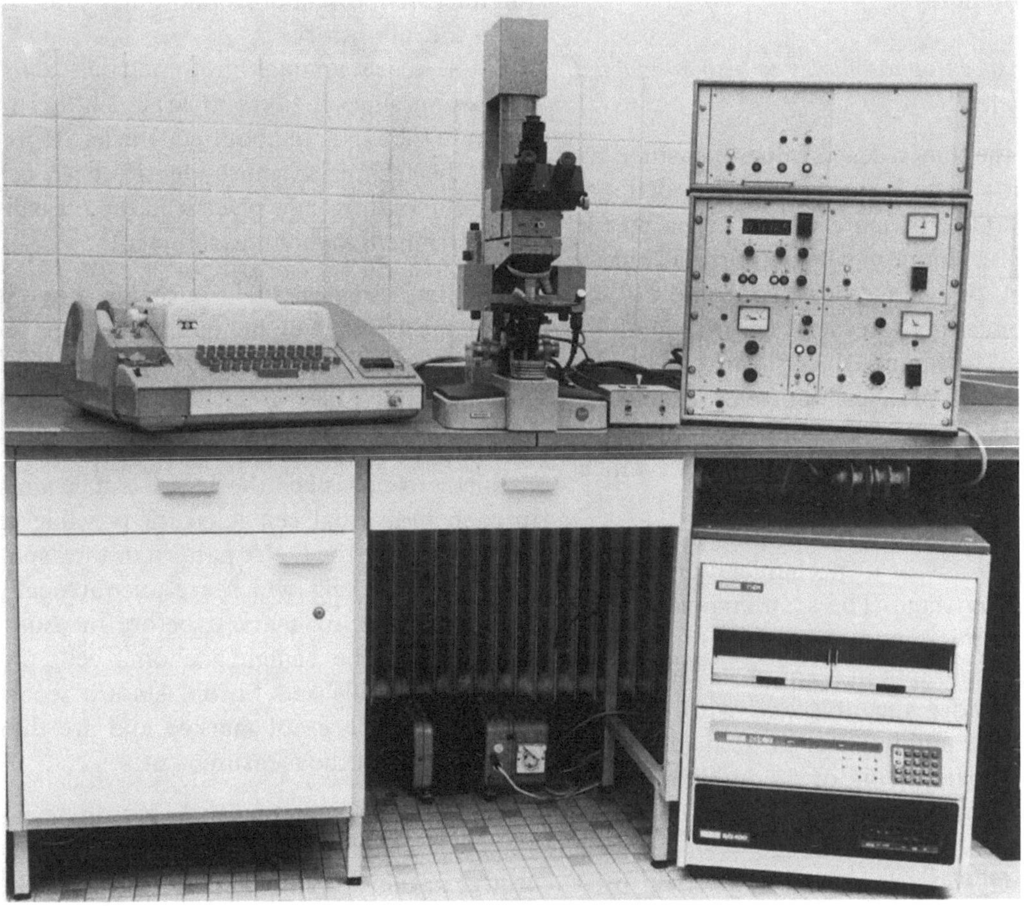

Fig. 229. Apparatus for single-cell scanning cytophotometry comprising microscope cytophotometer (*center*), cabinet for electronic parts (*right*), printer (*left*) and pdp 8a computer (manuf. by Digital) with floppy disk double drive (*below*)

13.2.2.1 Our own modified staining technique according to Feulgen

Preparation of Schiff's reagent. Shake 1 g fuchsin in 200 ml hot water, cool and filter; add 20 ml 1 N HCl and 1 g $K_2S_2O_5$ and shake. Allow to stand for 24 hours, them mix with 2 g activated charcoal and filter. Store in a refrigerator.
Schiff's reagent is also obtainable commercially as a ready-made product[1].

Fixation	
Dry in air	1 min
Carnoy's solution	15 min
Hydrolysis	
0.1 N HCl	1 min
5 N HCl	15–30 min (20 °C)
0.1 N HCl	1 min
Staining	
Schiff's reagent	60 min
Differentiate in SO_2 water (36 ml 10% sodium metabisulfite, 30 ml 1 N HCl, dilute to 600 ml with distilled water)	3 × 3 min
Rinse (running water)	3 × 5 min
Dehydrate (70%, 80%, 95% and 100% alcohol)	each 1 min
Xylene	1 min
Mount in Eukitt	1 min

Result: DNA is stained scarlet

The preparations can be stored for several months if protected from light

1 Manufactured by Merck, Darmstadt

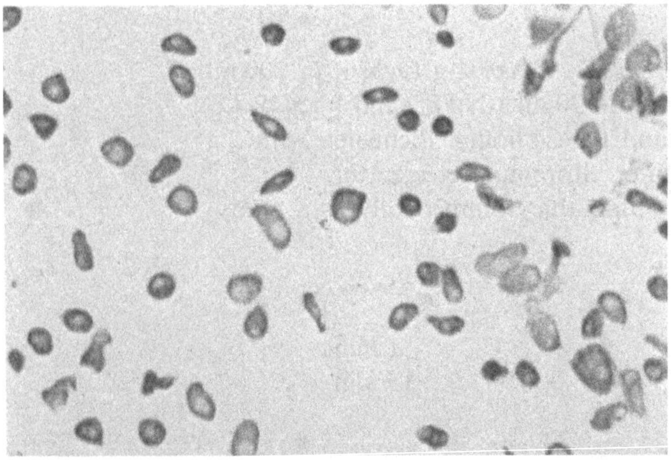

Fig. 230. Nuclei from a prostatic carcinoma stained scarlet with Feulgen's strain. ×400

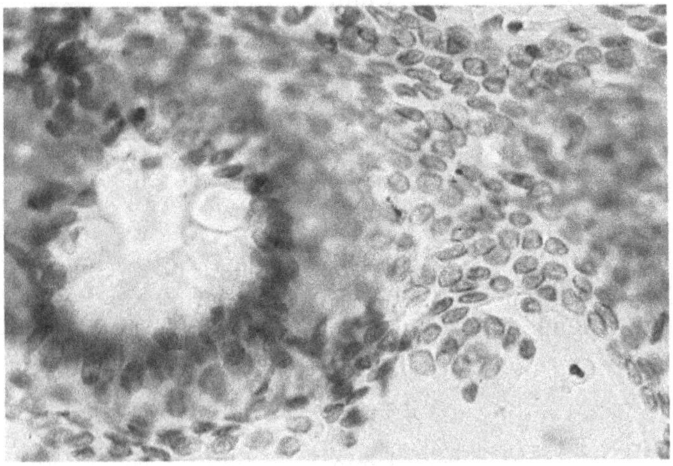

Fig. 231. Large group of epithelial cells from the rectal mucosa stained by Feulgen's technique. ×400

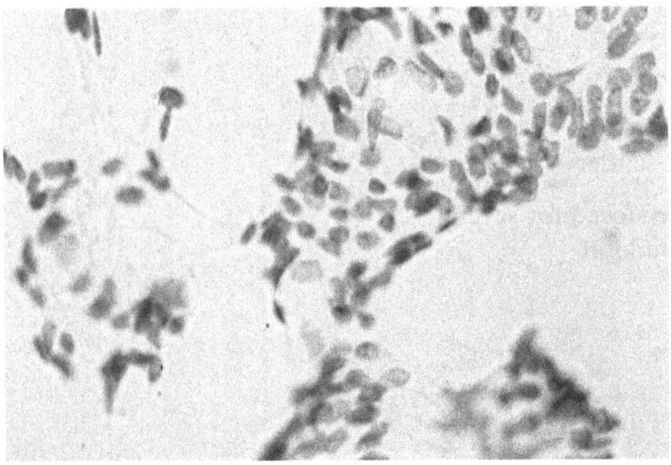

Fig. 232. Group of epithelial cells from the rectal mucosa stained by Feulgen's technique. ×400

Fig. 233. DNA cytophotogram with aneuploid DNA distribution and tetraploid DNA frequency peak

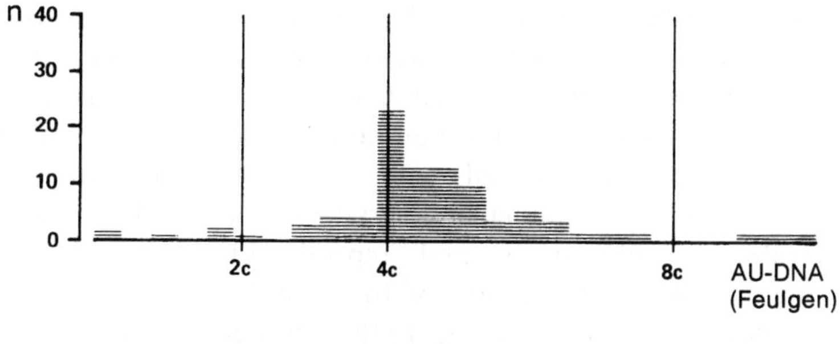

13.2.2.2 Possible Sources of Error

Fixing too long in Carnoy's solution. Fixation may on no account take longer than 15 minutes.

Schiff's reagent which has been stored too long stains the nuclear DNA pale red only or not at all as the fuchsin has lost the sulphuric acid. If the reagent is red due to free fuchsin it should no longer be used.

Exceeding the hydrolyzation time: If hydrolysis is carried out for too long this can result in the DNA's failing to become stained.

13.2.2.3 Re-staining After Staining with Papanicolaou's Stain

If aspirates have already been stained with Papanicolaou's stain for cytodiagnosis they can be decolorized after removal of the coverglasses and re-stained with Schiff's reagent (BÖCKING 1981). Thus older preparations can also be subjected to DNA cytophotometry.

Methods

Xylene (for removal of the coverglasses)	4 days
100% alcohol	10 min
70% alcohol	10 min
Distilled water	10 min

Hydrolysis followed by the typical Feulgen's reaction.

Hydrolysis may also be conducted in the dark at 28 °C with 4 N HCl for 45 minutes. Thereafter rinse the preparation with distilled water.

13.2.3 Cytophotogram

The DNA content of about 100 cell nuclei determined at each measurement is summarized in a cytophotogram and expressed as an absolute value in arbitrary units (AU). The number of cell nuclei measured by cytophotometry is given as "n" **(see Figs. 233, 239 and 240).**

The 2c value, i.e. the DNA content of a normal diploid set of chromosomes, is determined in *leucocytes* or *normal prostatic epithelium* as the standard diploid cell population.

These so-called standard cells (standard cell population) must be fixed in the same way as the prostate nuclei to be measured and hydrolyzed and stained with Schiff's reagent.

When the 2c value is doubled this is shown as 4c in the cytophotogram, meaning a tetraploid set of chromosomes, and a further doubling (quadrupling) gives an octaploid set of chromosomes (8c).

The DNA content of a nucleus is termed either *euploid* or *polyploid* if it is within the range of the mean value of the diploid standard cell population (±25%) or a multiple of it (±25%). All nuclei outside of the emploid or polyploid ranges are termed *aneuploid* (SPRENGER et al. 1974).

13.2.4 Statistics

The frequency distribution of the DNA content of tumor cells can exhibit widely varying characteristics, and thus it is hardly possible to make a purely descriptive assessment with adequate reliability, even though it is possible to distinguish between normal, non-proliferative and proliferative cells.

However, if the malignancy grade is to be established objectively and reproducibly, quantitative determination of the extent of the deviation from the normal DNA content and its frequency distribution (ploidy distribution) among the number of nuclei measured is absolutely necessary. When cytophotometry is combined with the appropriate methods of statistical analysis it is possible to distinguish accurately between proliferating and non-proliferating cell populations and the extent of proliferation.

Three different procedures can be used for the statistical calculation of the ploidy distribution:

- calculation of the 2c "deviation index" and the 4.5c "exceeding rate" (Böcking 1981).
- the Median Quartile Test (Bauer 1962; Seppelt and Sprenger 1981).
- a modified Median Quartile Test (Leistenschneider and Nagel 1983).

13.2.4.1 The 2c "Deviation Index" and the 4.5c "Exceeding Rate"

The 2c deviation index (2c-DI) gives the extent of deviation of the individual ploidies obtained from the diploid (2c) nuclear DNA content in squared unit weight (Böcking 1981). This index is thus a statistical parameter which represents a theoretical limit between benign and malignant nuclei.

However, the 2c-DI alone is not reliable as a criterion of malignancy since, if there is a high degree of proliferative activity or euploid polyploidization, i.e. higher powers of the diploid (2c) nuclear DNA content, the 2c deviation indices (2c-DI) are exaggerated, and thus false-positive diagnoses are possible if this parameter is used alone.

If a second statistical parameter is used, namely the 4.5c exceeding rate (4.5c-ER), which expresses the percentage of distinctly aneuploid nuclei, a more reliable diagnosis of malignancy can be made, and the nuclear DNA values obtained can be taken as a basis for DNA malignancy grading (Böcking 1981). Unlike the 2c-DI the 4.5c-ER is *not* raised by euploid polyploidization. Moreover, nuclei with a DNA content of between 2c and 4c, which can be in the S phase of the cell cycle but can also be cancerous, are not covered by the 4.5c-ER.

13.2.4.2 The Median Quartile Test

Using the quartiles of the individual DNA distributions per patient the Median Quartile Test for differences in location, dispersion and type of distribution gives the limits of the classes, which each contain 25% of these quartiles. The quartiles are calculated from the medians of each individual distribution of the values measured and the two quartiles to the right and left of the medians. The median is the point of separation which divides the distribution into two equal halves, each of which contains the same number of measurements.

The values for the DNA distribution differ significantly from quartile to quartile and thus it is possible to express different malignancy potentials by means of the quartiles (Seppelt and Sprenger 1981).

13.2.4.3 The Modified Median Quartile Test

Statistical classification of the values in the DNA frequency distribution is easier with this test than with the Median Quartile Test itself. It is based on the assertion of the Central Limit Theorem, namely, that the distribution of the mean values of *approximately* normally distributed values is *always* normal. Thus the sum of the mean values of each patient having undergone examination for the DNA content of prostatic cell nuclei is included in the calculation. For example, if 50 patients undergo examination the overall mean value is calculated from the 50 individual means. Then the 25% limits to the right and left are fixed (Z transformation of the normal distribution). In this way 3 classes can be formed which in accordance with the Gaussian distribution have 50% of the mean values in the centre and 25% in each of the "tails". The differences between them are tested for significance by analysis of variance using Student's t-test. If there is a significant difference between the 3 classes the various cytophotograms can be classified for malignancy accordingly, i.e. according to the DNA content (LEISTENSCHNEIDER and NAGEL 1983).

13.3 Flow-through Cytophotometry

In flow-through cytophotometry (impulse cytophotometry) the DNA is measured in suspended cells, which are directed along the microscope's optical path.

The cells are fluorochromated with acriflavine-Schiff's reagent or ethidium bromide and are then passed, in the suspension, through the sampling inlet into the measuring chamber, where they move through the range of the depth of focus of the microscope objective, parallel to the optical axis **(Fig. 234).** Here they are subjected to the maximum fluorescence excitation and emit a fluorescence signal proportional to the DNA content of the nucleus, which is amplified by a photomultiplier and stored in a computer. After measurement each individual cell is rinsed away by a stream of fluid. In this manner all fluorescence impulses emitted during a measuring procedure are recorded in terms of light intensity, and classified. The result of the measurement is expressed in a cytophotogram (histogram)

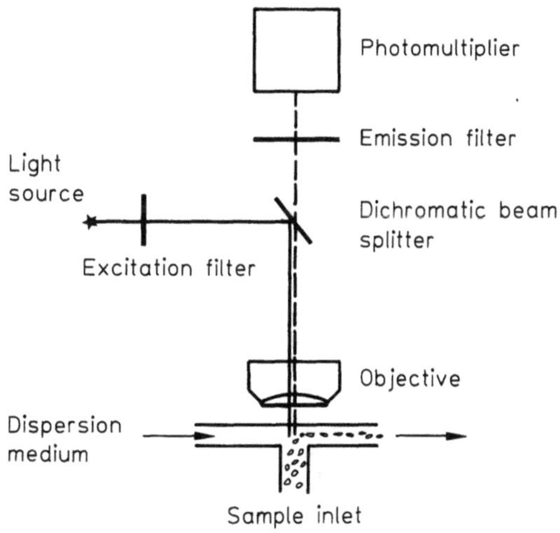

Fig. 234. Diagram of the optical section of an automatic flow-through cytofluorometer. The cell suspension is passed in parallel to the optical axis of the microscope and at right angles to the cell-free laminar flow, in accordance with the mechanical focussing principle. (von SENGBUSCH and HUGEMANN 1974)

showing the fluorescence intensities and thus the DNA contents of the nuclei **(see Fig. 235).**

The *advantage of flow-through cytophotometry*[1] as compared with single-cell scanning cytophotometry lies in the fact that a sample can be measured rapidly, the measuring procedure, including the evaluation of on average 20,000 cells, being concluded in about 15 minutes.

However, the decisive disadvantage of the method is that it is impossible to assign the DNA content measured to the morphology of the cell examined, and thus the results are not reproducible!

The reason why the values obtained cannot be assigned to malignant cells alone and the results of flow-through cytophotometry cannot be reproduced is that during flow-through cytophotometry not only the prostatic carcinoma cells in the aspirate are recorded but also all other fluorescing particles

1 Apparatus manufactured by Phywe, Göttingen, W. Germany and Ortho-Instruments, New York

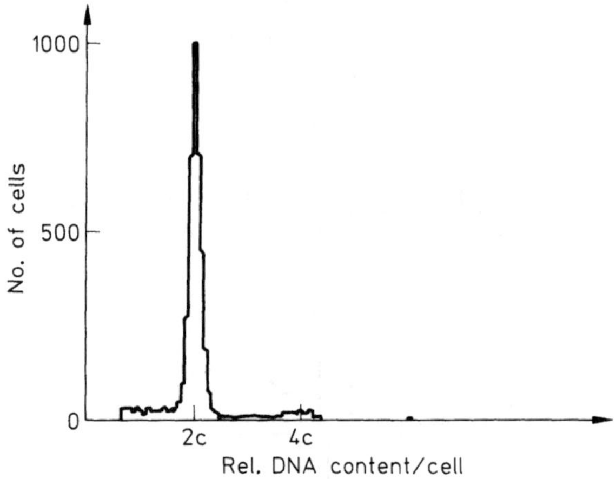

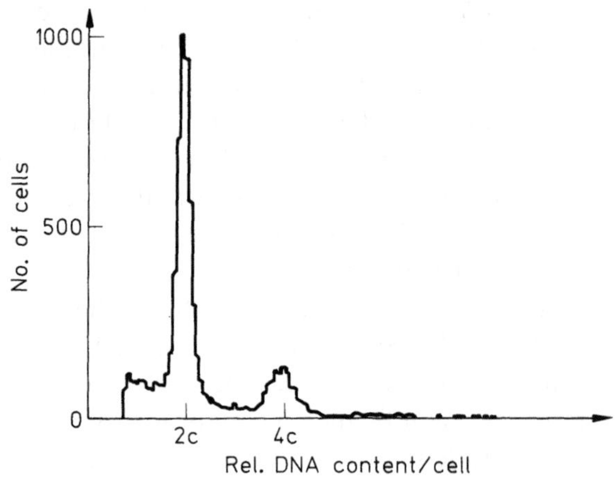

Fig. 235. Histograms of nuclear DNA in prostatic ademonas (*left*) and prostatic carcinomas (*right*) obtained with the aid of impulse cytophotometry. (ZIMMERMANN et al. 1979)

present in the sample, for example benign adenoma nuclei, nuclei of inflammatory cells, damaged cells, incompletely separated nuclei and fluorescing impurities. Thus cytophotograms obtained by means of flow-through cytophotometry can be considerably distorted. The presence of nuclei of activated cells, such as are found in inflammation, for example, is especially liable to lead to a considerable shift of the DNA distribution to the right and thus simulate carcinoma.

In practice this means that there is a danger of obtaining false-positive results.

On the other hand, false-negative findings can result from the presence of a relatively small proportion of atypical cells in a mixed population of atypical and normal cells or from carcinomas whose DNA frequency distribution is the same as that of normal tissue with a diploid DNA stemline (SPRENGER 1979).

The rate of false-positive results obtained for aspirates from prostatic tissue with flow-through cytophotometry is 30% and that of false-negative findings 11% (SPRENGER 1979). Errors of this kind can only be eliminated by subjecting the cells to a sorting procedure, however, none is as yet available for routine use **(Fig. 236).**

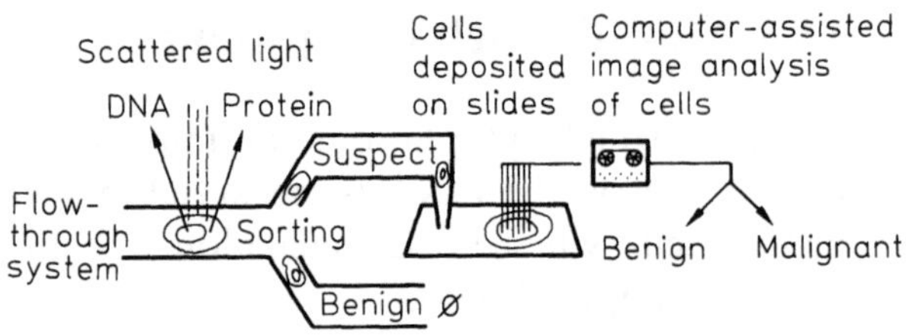

Fig. 236. Diagram of possible apparatus for a diagnostic system comprising flow-through cytophotometry and cell sorting. (SPRENGER 1979)

13.4 New Developments in Automatic Cytodiagnosis

13.4.1 The A.S.M. System

The purpose of automatic cytodiagnosis of prostatic carcinoma is to ensure recognition of carcinoma cells among the large number of different types of cells obtained by aspiration or punch biopsy. Determination of the DNA content of nuclei alone is not enough.

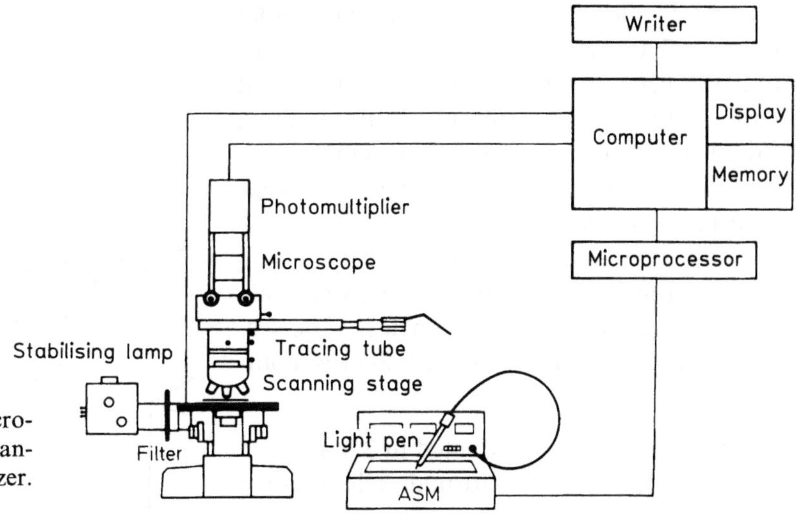

Fig. 237. Combination of a microscope cytophotometer with a manually operated image analyzer. (AEIKENS and LIEDTKE 1982)

It can only be achieved with an additional, reliable method of separating cells and nuclei with the aid of analytic systems with a high resolving power capable of detecting and recording a large number of relevant cell parameters in a *single* evaluation process.

These requirements are fulfilled by the A.S.M. analysis system[1] **(see Fig. 237).** With this system numerous characteristics such as nuclear surface area, cytoplasmic surface area, mean absorption of the cytoplasm, total absorption of the cytoplasm and the nuclear-cytoplasmic ratio can be quantitatively evaluated and automatically recognized.

In this system the photometer microscope is linked to a manually operated image analyzer. The two instruments are linked optically by means of a tracing tube attached to the photometer microscope. The cell under examination, the field and measuring field diaphragms of the photometer microscope and the light-emitting diode at the tip of the tracing pen of the manually operated image analyzer are visible through the binocular tube of the photometer microscope. Following single-cell cytophotometry the light pen is used to trace both the nucleus and the cytoplasm of the cell projected onto the image analyzer by the tracing tube, which are thus divided into segments so that quantitative analysis of the various parameters can be carried out automatically.

Whereas a high degree of differentiation with an error rate of only 5% has been obtained by automatic recognition and classification of *urothelial cells* into normal, atypical and malignant cells it has not yet been possible to achieve similarly good differentiation of tumor and other cells in *cytology of the prostate* (AEIKENS and LIEDTKE 1982).

13.4.2 The Leytas System

The aim of this system (*LEY*DEN *T*ELEVISION *A*NALYSIS *S*YSTEM) is to achieve both automatic differential diagnosis of prostatic adenoma and carcinoma by combining flow-through cytophotometry and cell sorting with automatic texture analysis (TAS[1]).

Flow-through cytophotometry and cell sorting are carried out with violet light using

1 Manufactured by Leitz, Wetzlar

an FACS[1] flow sorter **(Fig. 238).** The violet light (457 nm) is excited by an argon laser[2] and the cells are in suspension and stained with mithramycin-ethidium. Cells with a raised DNA content are then removed in centrifuge chambers for image analysis (TANKE et al. 1982).

The Leytas system consists of a special version of the TAS system and a gray scale memory for storing cell parameters. The system is linked digitally to a PDP-1104 computer[3].

Image analysis is conducted in 3 stages:

- rapid selection of abnormal cells, storage in a gray scale memory;
- interactive phase with visual examination of the stored cells;
- elimination of remaining artefacts and quantitative analysis of various nuclear parameters, e.g. chromatin.

A special program can be used to recognize abnormal cells on the basis of their raised DNA content or density of chromatin.

The sensitivity of the Leytas system in the primary diagnosis of prostatic carcinoma is at present still low. Positive results were obtained in only 10 out of 23 carcinomas (43%). In 6 cases the material was negative (26%) and in 7 cases insufficient (30%) (dE VOOGT et al. 1981).

1 Manufactured by Becton-Dickinson (Sunnyvale, Ca., USA)

2 Manufactured by Spectra Physics (Mountain View, Ca., USA)

3 Manufactured by Leitz (Wetzlar)

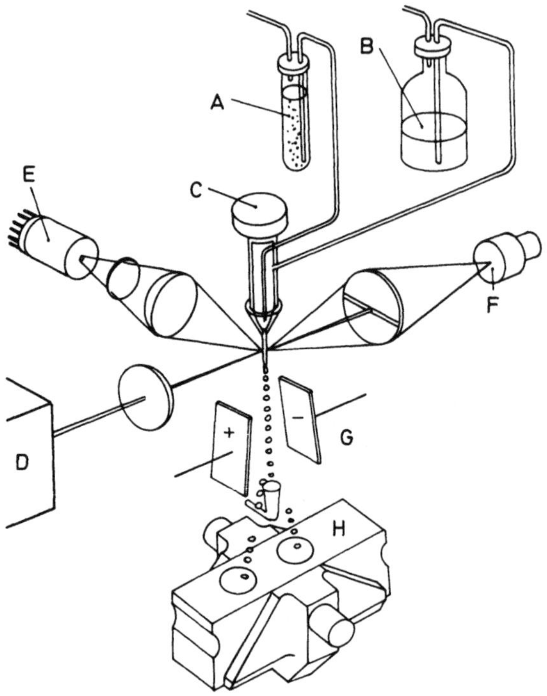

Fig. 238. Diagram of a flow-through sorting instrument for sorting cells for further image analysis. *A*, cell suspension; *B*, saline; *C*, pipette with vibration element; *D*, laser equipment; *E*, fluorescence detector; *F*, scattered light detector; *G*, sorting electrodes; *H*, centrifuge chamber. (TANKE et al. 1982)

However, on the whole the Leytas system is a combination with which substantially more reliable automatic cytodiagnosis will be possible in future.

The reliability and hence the clinical value of the automatic cytological procedures described is always dependent on the provision of sufficient, representative cellular material from the prostate and thus on the mastery of the technique of aspiration.

14 Results of Measurement of Nuclear DNA by Single-cell Scanning Cytophotometry in Prostatic Carcinoma

14.1 Well Differentiated Carcinoma (Grade I)

This type of carcinoma does not exhibit a statistically significant increase in the middle ploidy range and thus the cytogram shows mainly the same DNA frequency distributions as those for nuclei with normal prostatic epithelium or prostatic adenomas **(Fig. 239),** with a slim peak in the diploid range (2c) (BÖCKING 1981; SEPPELT and SPRENGER 1981).

However, from time to time a well differentiated carcinoma will exhibit accumulations of numerous values beyond 2c (ZETTERBERG and ESPOSTI 1976), which are a manifestation of the sometimes striking tendency of these carcinomas to progress. The evidently variable biological behavior of the well differentiated carcinomas is accounted for in the grading system of the *Uropathological Study Group on Prostatic Carcinoma* by the "score" given at the primary diagnosis **(Tables 12 and 13**, pp. 82, 83).

Statistically the well differentiated prostatic carcinomas almost always differ significantly from the Grade II and III carcinomas as regards the DNA content of the nuclei (SEPPELT and SPRENGER 1981).

14.2 Moderately Differentiated Carcinoma (Grade II)

Carcinomas in this group exhibit highly varying DNA frequency patterns with peaks in the diploid, triploid, tetraploid or aneuploid ranges **(Fig. 240).** As compared with the highly differentiated carcinomas (Grade I) there is usually a significant difference in the middle ploidy range, while there is no significant difference from the dedifferentiated carcinoma, either in descriptive or statistical terms (ZETTERBERG and ESPOSTI 1976; LEISTENSCHNEIDER and NAGEL 1980; SEPPELT and SPRENGER 1981).

ZETTERBERG and ESPOSTI (1976) describe 2 groups of DNA distributions in moderately differentiated prostatic carcinomas (G II):

One group with a DNA frequency peak in the diploid range, as in prostatic adenoma and most highly differentiated prostatic carcinomas, and a second with an abnormally raised nuclear DNA content with aneuploidy and scattering of the values from the hyperdiploid to the pentaploid range.

These findings also indicate that the G II carcinomas have widely varying malignancy potentials.

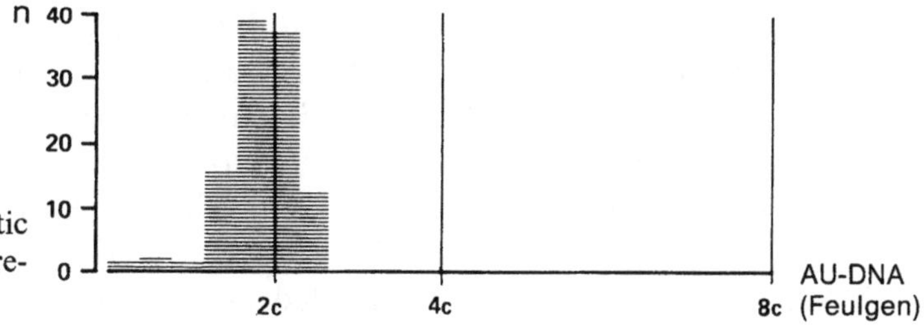

Fig. 239. DNA cytophotogram for prostatic carcinoma with a slim diploid (2c) DNA frequency peak

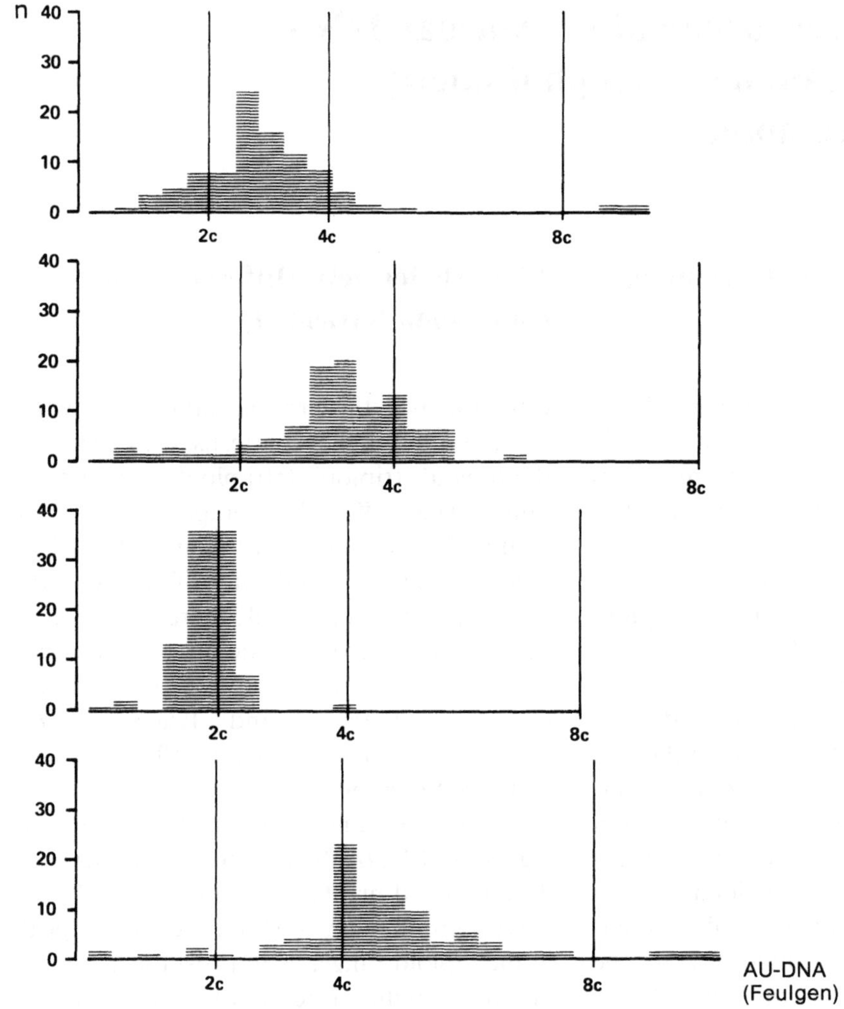

Fig. 240. DNA cytophotograms for moderately differentiated prostatic carcinomas with highly varying DNA distributions. (LEISTENSCHNEIDER and NAGEL 1980)

14.3 Undifferentiated Carcinoma (Grade III)

Wide variation of the values is typical. High, slim DNA frequency peaks in the 2c range are never found, only isolated diploid values (ZETTERBERG and ESPOSTI 1976; LEISTENSCHNEIDER and NAGEL 1979, 1980, 1983). The often extraordinarily great differences in DNA distribution in undifferentiated carcinomas also indicate the existence of highly heterogeneous types within this group.

In 15 patients with undifferentiated, treatment-resistant carcinomas we found a DNA peak at 3c or 4c in 13 cases which were assigned to Class II with the aid of the modified median quartile test. With values in the hypertetraploid range and scattered values higher than 8c, however, after statistical analysis these cases were re-assigned to Class III, which differs significantly from Class II (LEISTENSCHNEIDER and NAGEL 1983).

14.4 Our own Results with Nuclear DNA Analysis by Single-cell Cytophotometry in Treated Prostatic Carcinoma

In a prospective clinical study, analysis of nuclei was carried out in 20 patients with undifferentiated prostatic carcinoma (G III) over periods ranging from one week to a maximum of one year after the start of treatment (LEISTENSCHNEIDER and NAGEL 1983). A total of 67 nuclear DNA analyses were conducted, about 100 nuclei per patient being measured.

The following treatments were employed:

Primary estrogen therapy	2 patients
Primary cyproterone acetate therapy	3 patients
Primary therapy with estramustine phosphate	4 patients
Secondary therapy with estramustine phosphate	6 patients
Secondary therapy with Cyclophosphamide	1 patient
Tertiary therapy with Cyclophosphamide	6 patients

Before primary therapy was initiated each patient was subjected to a thorough examination by the conventional clinical methods, including bone scintigraphy and computer tomography.

The median quartile test (see p. 206) was used for *statistical evaluation.*

14.4.1 Nuclear DNA Distribution Patterns During Treatment

Irrespective of the form of treatment employed 3 different patterns of DNA distribution in the nuclei were found:

a fall from the aneuploid or polyploid to the diploid range with a statistically significant change from Class III or II to Class I **(Figs. 241–243);**

no definite changes in the DNA cytophotogram and a shift from Class III or II to Class I which was *not* statistically significant **(Fig. 244);**

a shift in the DNA frequency peak from the aneuploid or tetraploid to the octaploid range or beyond, with a statistically significant change from Class I to Class II or even III **(Figs. 245 and 246).**

14.4.2 Results

4 patients with a significant reduction in nuclear DNA content from the tetraploid or aneuploid to the diploid range (2c) no later than 12 weeks after the start of treatment, hence statistically coming under Class I, survived for an average of 2 years 5 months on primary therapy.

4 other patients ($T_3N_XM_0$) also with DNA frequency peaks at 2c after 12 weeks have so far survived for $3^1/_2$ years.

3 patients ($T_3N_XM_1$) on secondary or tertiary treatment not exhibiting a shift of the DNA frequency peak to 2c or rather, as in 2 cases, a shift towards the right of the DNA cytophotogram, had mean survival times of only 8 months.

Therefore a shift of the nuclear DNA content into the diploid range during treatment may generally be assessed as therapeutic success.

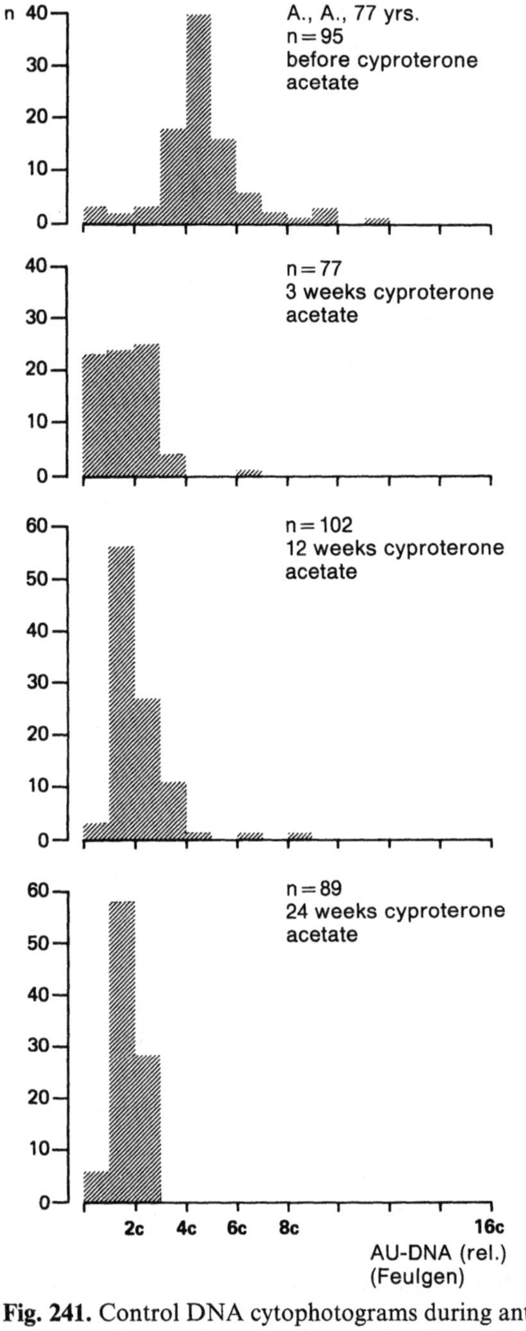

Fig. 241. Control DNA cytophotograms during antiandrogen therapy: 3 weeks after starting therapy, the DNA frequency peak is already near 2c, with numerous values in the hypodiploid (1c-) range. This shift to the left is statistically significant. In the further course of therapy, the peak remains in the diploid range.
The clinical classification of the carcinoma was grade III, stage $T_3N_xM_0$

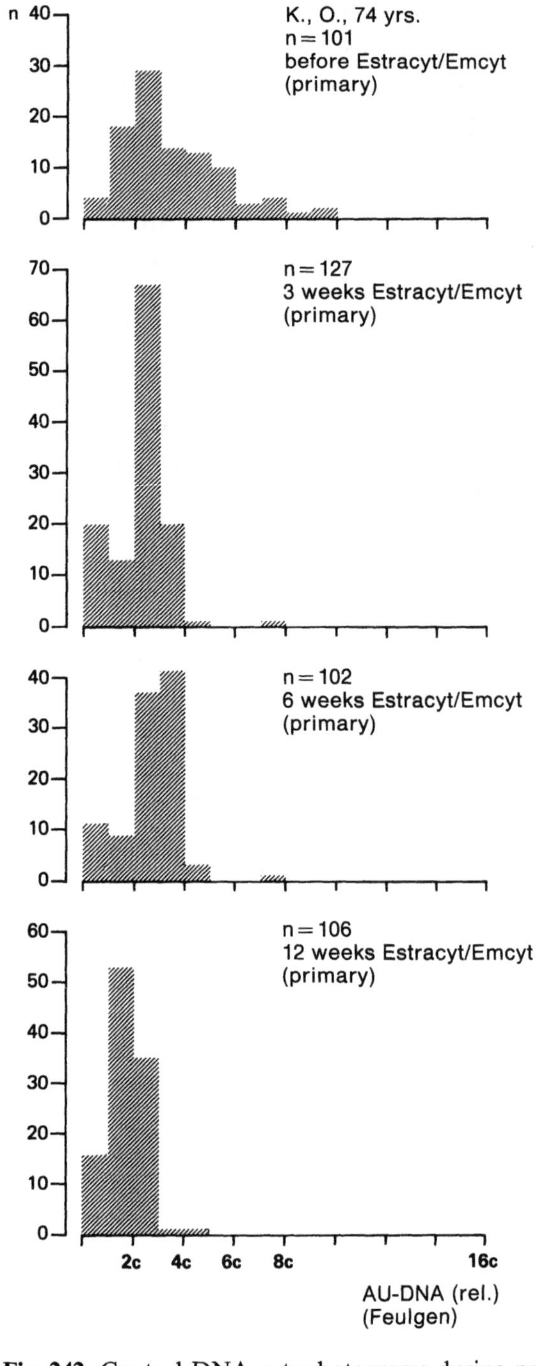

Fig. 242. Control DNA cytophotograms during primary *Estracyt* therapy: after 12 weeks of therapy, there is a slim (statistically significant) DNA frequency peak near 2c.
The clinical classification of the carcinoma was grade III, stage $T_3N_xM_0$. The patient has been clinically stable for 5 years

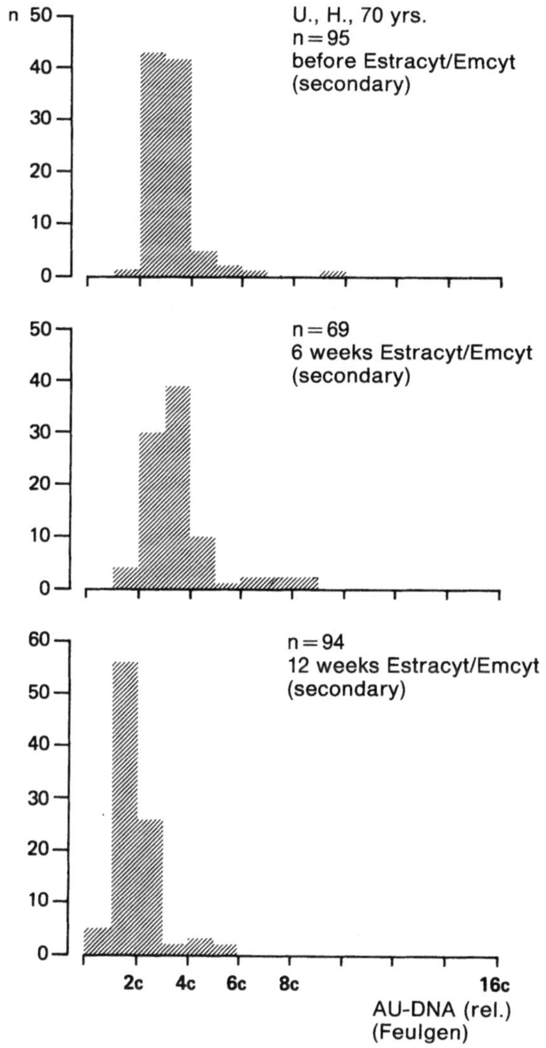

Fig. 243. Control DNA cytophotograms during secondary *Estracyt* therapy following previous hormone resistance: after 12 weeks of therapy, there is a statistically significant shift of nuclear DNA distribution to the left, with the peak near 2c.
The carcinoma was clinically classified as grade III, stage $T_3N_xM_0$. The patient has been clinically stable for $3^1/_2$ years

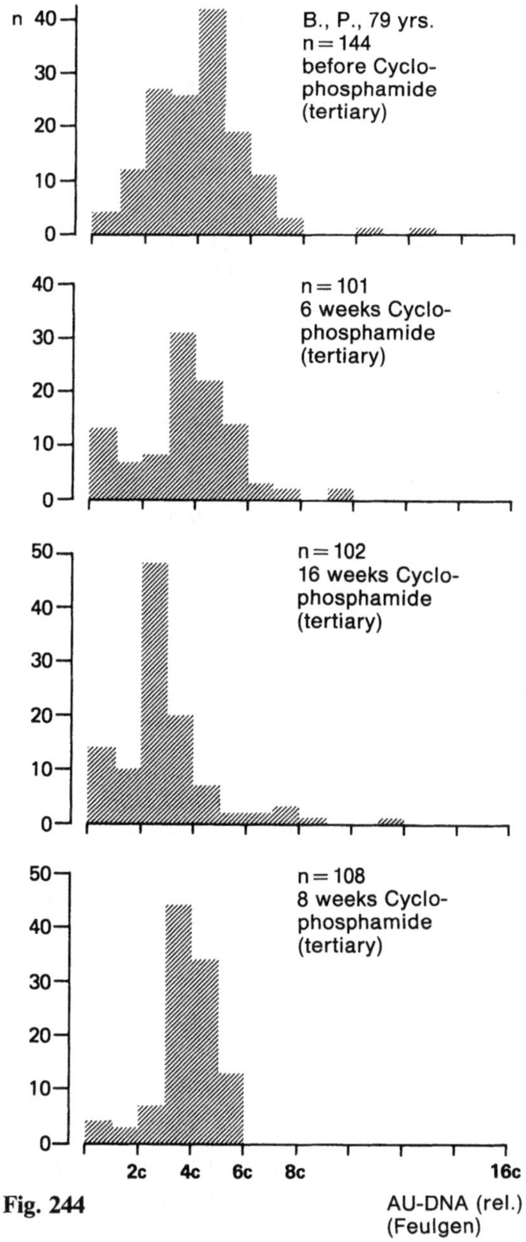

Fig. 244

Fig. 244. Control DNA cytophotograms during tertiary *Cyclophosphamide* therapy after resistance to primary hormone and secondary *Estracyt* therapy: prior to therapy, the tumor is aneuploid, with values scattered up to 8c. After 6 weeks of therapy, there is practically no change. After 8 weeks of therapy, there is a slim DNA frequency peak near 2c but numerous values still reach 6c, and some even beyond. Statistically there is no significant difference from the DNA distribution before therapy.
After 16 weeks of therapy, the peak is in the tetraploid range, with numerous values scattering to 6c, i.e. continuing aneuploidy. Prior to *Cyclophosphamide* therapy, the clinical classification of the carcinoma was stage $T_4N_2M_0$. After 16 weeks of therapy, clinical examination already revealed bone metastases

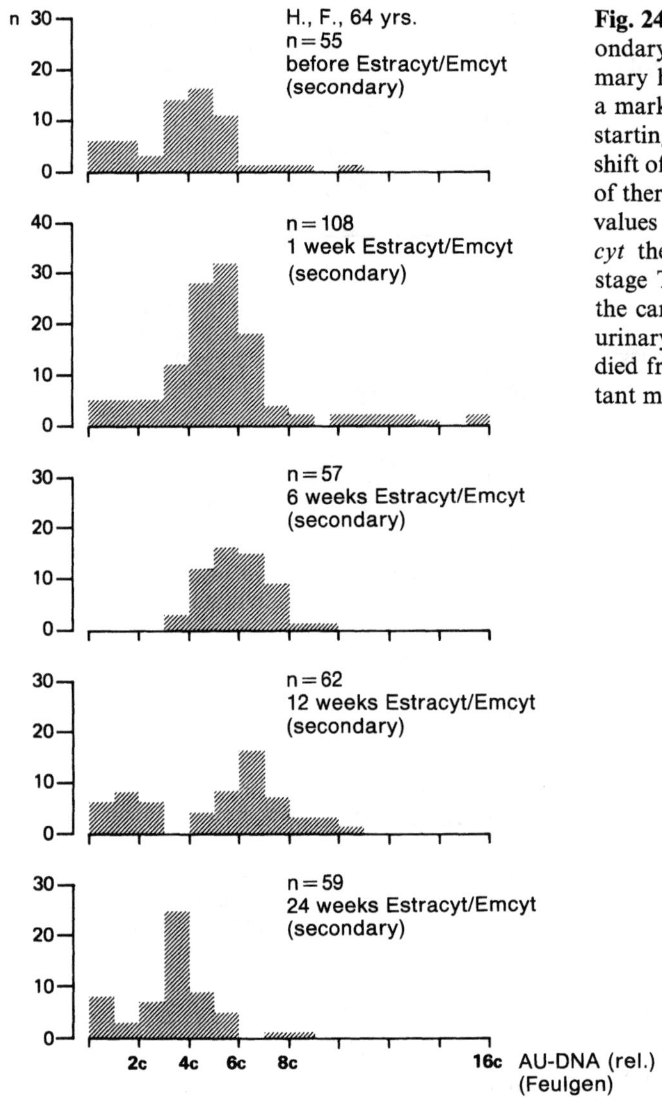

Fig. 245. Control DNA cytophotograms during secondary *Estracyt* therapy following resistance to primary hormone treatment: prior to therapy, there is a marked aneuploidy. Between 1 and 12 weeks after starting therapy, there is a statistically significant shift of DNA distribution to the right. After 24 weeks of therapy, the DNA frequency peak is near 4c, with values scattering up to 6c and beyond. Before *Estracyt* therapy, the tumor was clinically classified as stage $T_3N_1M_0$. After 24 weeks of *Estracyt* therapy, the carcinoma had already broken through into the urinary bladder (T_4), and 8 months later, the patient died from the primary disease due to increasing distant metastases

In contrast, treatment-resistant carcinomas exhibit DNA distribution patterns which may be considered typical of malignant proliferating tumors. Treatment-resistant, undifferentiated prostatic carcinomas show the same DNA distribution patterns as untreated prostatic carcinomas.

Undifferentiated, hormone-resistant carcinomas are *not* statistically different from carcinomas resistant to radiotherapy or treatment with estramustine phosphate or cytostatic drugs.

When treatment is successful the values are strikingly often found in the hypoploid range. These are probably the values for nuclei with DNA fragments and not those for haploid nuclei. It is possible that when tumor cell nuclei are arrested in individual phases of the cell cycle as a result of the treatment, e.g. in the G-I phase, there is an increase in depolymerase activity and thus a loss of DNA, which is initially only partial but later complete. It is, however, also conceivable that the nuclei damaged by the treatment

Fig. 246. Control DNA cytophotograms during tertiary *Cyclophosphamide* therapy following resistance to primary hormone and secondary *Estracyt* therapy: at 24 weeks after starting therapy, there is no statistically significant change of aneuploidy despite a slim DNA frequency peak near 2c. However, after 52 weeks of therapy, there is a massive shift of DNA distribution to the right. Prior to *Cyclophosphamide* therapy, the carcinoma was clinically classified as stage $T_4N_2M_0$. After 24 weeks of therapy, bone metastases were already detected which, after 52 weeks, had spread diffusely. Following cross-over to 5-fluoro-uracil therapy, the patient survived on cytostatic therapy for $2^1/_2$ years before he died of the carcinoma

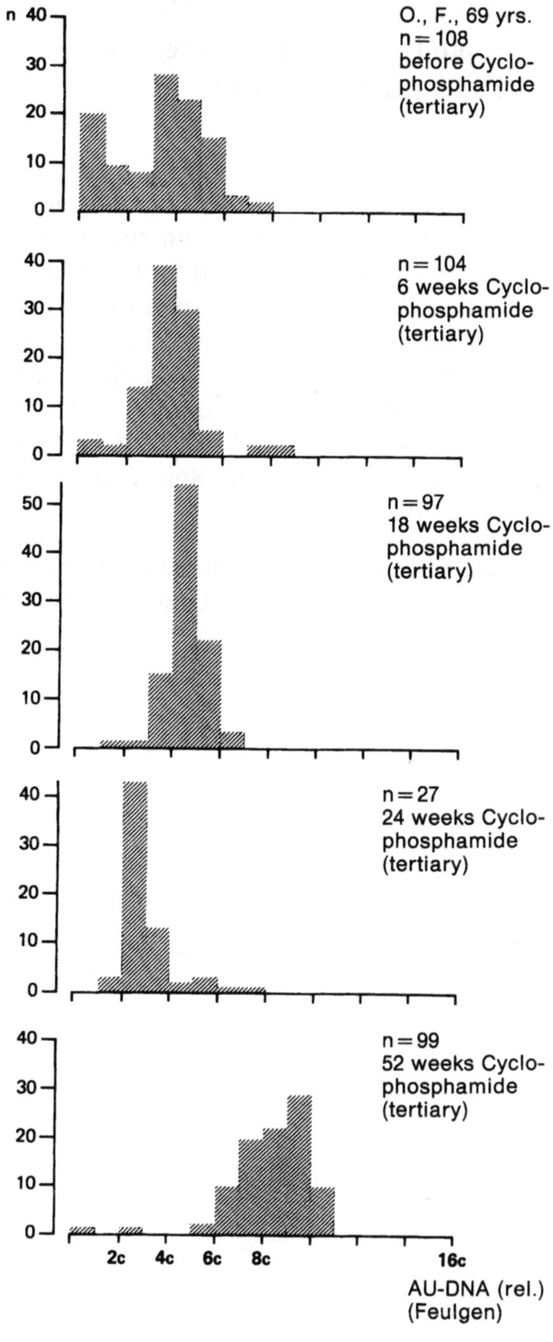

fragment on further processing and that this results in a hypodiploid DNA content, in other words, these are artefacts.

In terms of the cell cycle when DNA frequency peaks shift to the diploid range during treatment this means that the majority of the carcinoma nuclei are in the G-I phase. This pattern of DNA distribution is also the same as that of prostatic adenomas. It is likely that the nuclei have been arrested in the G-I phase and thus progression from the G-I phase to the S phase is prevented. This also explains why a clinical course with drastic, statistically significant reduction of the DNA frequency distribution is more favorable, whereas conversely, rapid progression is to be expected if the DNA content rises.

14.5 The Importance of DNA Cytophotometry in the Treatment of Prostatic Carcinoma

It would appear that the investigation of the effect of a given treatment on the primary tumor itself is very important for the prognosis of the biological behavior of a tumor during a given form of treatment. The following conclusions can be drawn from the results of nuclear DNA analysis obtained to date (LEISTENSCHNEIDER and NAGEL 1979, 1980, 1982, 1983; SEPPELT and SPRENGER 1981; BÖCKING 1981):

- well differentiated carcinomas (Grade I) differ significantly from Grade II and III carcinomas as regards the DNA content of their cell nuclei;
- moderately differentiated carcinomas frequently do *not* differ significantly from undifferentiated carcinomas (Grade III) as regards nuclear DNA content;
- treatment-resistant carcinomas have a significantly higher nuclear DNA content than successfully treated carcinomas;
- a significant fall in nuclear DNA content generally indicates that the tumor is responding to the treatment and that the prognosis is good;
- there is a correlation between a persistently atypical nuclear DNA content and poor prognosis.

References

Ackermann R, Müller HA (1977) Retrospective analysis of 645 simultaneous perineal punch biopsies and transrectal aspiration biopsies for diagnosis of prostatic carcinoma. Europ Urol 3:29

Aeikens B, Liedtke GE (1982) Untersuchungen zur Meßbarkeit und Objektivierbarkeit der Urinzytologie durch ein hochauflösendes mikroskopzytophotometrisches Meßsystem. Urologe A 21:98

Alken CE, Dhom G, Straube W, Braun JS, Kopper B, Rehker H (1975) Therapie des Prostata-Carcinoms und Verlaufskontrolle (III). Urologe A 14:112

Arduino LJ, Murphy FJ (1963) Carcinoma of the prostate: Use of the Franzén needle for perineal prostatic biopsy. J Urol 89:732

Barnes RW, Ninan CA (1972) Carcinoma of the prostate: Biopsy and conservative therapy. J Urol 108:897

Bertelsen S (1966) Transrectal needle biopsy of the prostate. Acta Chir Scand 226:357 (Suppl)

Bichel P, Frederiksen P, Kjaer T, Thommesen P, Vindeløv LL (1977) Flow microfluorometry and transrectal fine needle biopsy in the classification of human prostatic carcinoma. Cancer 40:1206

Bishop D, Oliver JA (1977) A study of transrectal aspiration biopsies of the prostate with particular regard to prognostic evaluation. J Urol 117:313

Bissada NK (1977) Accuracy of transurethral resection of the prostate versus transrectal biopsy in the diagnosis of prostatic carcinoma. J Urol 118:61

Böcking A (1980) Grading des Prostatakarzinoms. Habilitationsschrift, Freiburg 1980

Böcking A (1981) Reproduzierbares zytologisches Malignitätsgrading des Prostatakarzinoms. Akt Urol 12:278

Böcking A, Sommerkamp H (1980) Histologisches Malignitätsgrading des Prostatakarzinoms. Prognostische Validität, Reproduzierbarkeit und Repräsentativität. Verhdlg Dtsch Ges Urol 32:63

Böcking A, Kiehn J, Heinzel-Wach M (1982) Combined histological grading of prostatic carcinoma. Cancer 50:288

Bourne CW, Frishette WA (1967) Prostatic fluid analysis and prostatitis. J Urol 97:140

Burck HC (1966) Histologische Technik. Leitfaden für die Herstellung mikroskopischer Präparate in Unterricht und Praxis. Thieme, Stuttgart

Byar DP, Mostofi FK (1972) Carcinoma of the prostate: Prognostic evaluation of certain pathologic features in 208 radical prostatectomies, examined by the stepsection technique. Cancer 30:5

Cartagena R, Baumgartner G, Wajsman Z, Merrin C (1975) Primary reticulum cell sarcoma of the prostate gland. Urology 5:815

Caspersson T (1936) Über den chemischen Aufbau der Strukturen des Zellkerns. Scand Arch Physiol 73:8

Chiari R, Harzmann R (1975) Perineale und transrektale Stanzbiopsie der Prostata. Urologe A 14:296

Cosgrove MD, George FW, Tery R (1973) The effects of treatment on the local lesion of carcinoma of the prostate. J Urol 109:861

Dajani YF, Burke M (1976) Leukemic infiltration of the prostate: A case study and clinocopathological review. Cancer 38:2442

Daves JA, Tomskey GC, Cohen AE (1961) Transrectal needle biopsy of the prostate. J Urol 85:180

Davison P, Malament M (1971) Urinary contamination as a result of transrectal biopsy of the prostate. J Urol 105:545

deVoogt HJ, Ploem JS, Brussee JAM, Knepfle C (1981) Verbesserung der zytologischen Diagnose beim Prostatakarzinom mit Hilfe des Leydener Fernseh-Analysesystems (Leytas). Verhdlg Dtsch Ges Urol. 32:66

Determann H, Lepusch F Das Mikroskop und seine Anwendung. Ernst Leitz GmbH, Wetzlar

Dhom G (1980) Pathologie des Prostatakarzinoms. Verhdlg Dtsch Ges Urol 32:9

Droese M, Soost H-J, Voeth C (1976) Zytodiagnostik des Prostatakarzinoms nach transrektaler Saugbiopsie. Urologe A 15:13

Editorial of Acta Cytologica (1964) The meaning of class 3 of the Papanicolaou classification of specimens from the female genital tract. Acta Cytol 8:99

Egle N, Spieler P, Bandhauer K, Gloor F (1976) Die Bedeutung zytologischer Untersuchungen für die primäre Diagnostik und den klinischen Verlauf des Prostatakarzinoms. Akt Urol 7:355

Ekman H, Hedberg K, Persson PS (1967) Cytological versus histological examination of needle biopsy specimens in the diagnosis of prostatic cancer. Brit J Urol 39:544

Emmett JL, Barber KW jr, Jackman RJ (1962) Transrectal biopsy to detect prostatic carcinoma: A review and report of 203 cases. J Urol 87:460

Epstein NA (1976) Prostatic carcinoma. Correlation of histologic features of prognostic value with cytomorphology. Cancer 38:2071

Epstein NA (1976) Prostatic biopsy. A morphologic correlation of aspiration cytology with needle biopsy histology. Cancer 38:2078

Esposti PL (1966) Cytologic diagnosis of prostatic tumors with the aid of transrectal aspiration biopsy. A critical review of 1100 cases and a report of morphologic and cytochemical studies. Acta Cytol 10:182

Esposti PL (1971) Cytologic malignancy grading of prostatic carcinoma by transrectal aspiration biopsy. Scand J Urol Nephrol 5:199

Esposti PL (1982) Aspiration biopsy and cytological evaluation for primary diagnosis and follow-up. In: Jacobi GH, Hohenfellner R Prostate cancer. Williams and Wilkins, Baltimore London

Esposti PL, Franzén S, Zajicek J (1968) The aspiration biopsy smear. In: Koss LG, Lippincott JB Diagnostic cytology and its histopathologic bases, 2nd Ed, Philadelphia

Esposti PL, Elman A, Norlén H (1975) Complications of transrectal aspiration biopsy of the prostate. Scand J Urol Nephrol 9:208

Faul P (1974) Die klinische Bedeutung der Prostatazytologie und ihre diagnostischen Möglichkeiten. Münch Med Wschr 116:15

Faul P (1975) Prostata-Zytologie. Steinkopf, Darmstadt

Faul P (1980) Vortrag auf der 1. Sitzung des Pathologisch-Urologischen Arbeitskreises „Prostatakarzinom". Bonn, 22.2.1980

Faul P, Schmiedt E, Kern R (1978) Die prognostische Bedeutung des zytologischen Differenzierungsgrades beim östrogenbehandelten Prostata-Carcinom. Urologe A 17:377

Feulgen R, Rossenbeck H (1924) Mikroskopisch-chemischer Nachweis einer Nukleinsäure vom Typ Thymusnukleinsäure und die darauf beruhende elektive Färbung von Zellkernen in mikroskopischen Präparaten. Z Physiol Chem 135:203

Ferguson RS (1930) Prostatic neoplasms: Their diagnosis by needle puncture and aspiration. Amer J Surg 9:507

Ferguson RS (1937) Diagnosis and treatment of early carcinoma of the prostate. J Urol 37:774

Fortunoff S (1962) Needle biopsy of the prostate: A review of 346 biopsies. J Urol 87:159

Frank IN, Scott WW (1958) The cytodiagnosis of prostatic carcinoma: A follow-up-study. J Urol 79:983

Franzén S, Giertz G, Zajicek J (1960) Cytological diagnosis of prostatic tumours by transrectal aspiration biopsy: a preliminary report. Brit J Urol 32:193

Gaeta JF (1981) Glandular profiles and cellular patterns in prostatic cancer grading. Urology 17:33 (Suppl)

Gaetani CF, Trentini GP (1978) Atypical hyperplasia of the prostate. A pitfall in the cytologic diagnosis of carcinoma. Acta Cytol 22:483

Garrett M, Jassie M (1976) Cytologic examination of post prostatic massage specimens as an aid in diagnosis of carcinoma of the prostate. Acta Cytol 20:126

Gleason DF (1966) Classification of prostatic carcinomas. Cancer Chemother Rep 50:125

Grayhack JT, Bockrath JM (1982) Diagnosis of carcinoma of the prostate. Urology 17:54 (Suppl)

Harada M, Mostofi FK, Corle DK, Byar DP, Trump BF (1977) Preliminary studies of histologic prognosis in cancer of the prostate. Cancer Chemother Rep 61:223

Hedinger Ch (1979) Pathologische Anatomie der Hodentumoren. Verhdlg Dtsch Ges Urol 30:7

Hell K, Graber P, Petronic V (1971) Die perineale Biopsie der Prostata mit Franklin-Vim-Silvermann-Nadel. Akt Urol 2:3

Helpap B (1980) The biological significance of atypical hyperplasia of the prostate. Virchows Arch A Anat Histol 387:307

Hohbach Ch, Dhom G (1978) Experiences with the prostatic cancer registry. 2nd Int Symp on the treatment of carcinoma of the prostate. Berlin, 1978

Hohbach Ch, Kopper B, Reichert HE, Dhom G (1980) Zur Pathologie der granulomatösen Prostatitis. Verhdlg Dtsch Ges Urol 31:39

International Union against Cancer (UICC) (1979) TNM; Klassifikation der malignen Tumoren. 3. Auflage. Springer, Berlin Heidelberg New York

Johnson DE, Chalbaud R, Ayala AG (1974) Secondary tumors of the prostate. J Urol 112:507

Kastendieck H (1980) Correlations between atypical primary hyperplasia and carcinoma of the prostate. Path Res Pract 169:367

Kastendieck H, Altenähr E, Geister H (1974) Das

Myosarkom der Prostata. Dtsch Med Wschr 99:392
Kastendieck H, Altenähr E, Burchardt P (1975) Zur Ultrastruktur des behandelten Prostatacarcinoms. Befunde nach kombinierter Hormon- und Strahlentherapie. Virch Arch Abt A Path Anat und Histol 366:287
Kastendieck H, Altenähr E (1975) Morphogenese und Bedeutung von Epithelmetaplasien in der menschlichen Prostata. Eine elektronen-mikroskopische Studie. Virch Arch Abt A Path Anat und Histol 365:137
Kastendieck H, Altenähr E, Burchardt P, Becker H, Franke HD, Klosterhalfen H (1976a) Morphologische und klinische Behandlungsergebnisse nach kombinierter Hormon- und Strahlentherapie des Prostatakarzinoms. Dtsch Med Wschr 101:571
Kastendieck H, Altenähr E, Husselmann H, Bressel M (1976b) Carcinoma and dysplastic lesions of the prostate. A histomorphological analysis of 50 total prostatectomies by step-section technique. Z Krebsforsch 88:33
Kaufman JJ, Schultz JI (1962) Needle biopsy of the prostate: A re-evaluation. J Urol 87:164
Kaufman JJ, Ljung BM, Walter Ph, Wajsman J (1982) Aspiration biopsy of prostate. Urology 19:587
Kaulen H, Davidts HH, Albrecht KF, Liebegott G (1973) Die besonderen Vorteile der Aspirationszytologie beim Prostata-Carcinom aus klinischer Sicht. Verhdlg Dtsch Ges Path 57:313
Keller AJ, Völter D, Bertsch H (1981) Die Wertigkeit zytomorphologischer Parameter unter der Hormontherapie des Prostata-Karzinoms. Urologe A 20:228
Khan O, Pearse E, Williams G (1982) Prostatic aspiration cytology. Its value in diagnosis and monitoring of response to treatment in patients with carcinoma of the prostate. 5. Kongreß der Europäischen Vereinigung für Urologie, Wien, 12.–15.5.1982
Koch KF (1972) Fluoreszenzmikroskopie. Instrumente, Methoden, Anwendung. Ernst Leitz GmbH, Wetzlar
Köllermann MW, Pessow D, Wagenknecht LV (1975) Komplikationen nach transrektaler Prostatabiopsie. Urologe B 15:225
Kohnen PW, Drach GW (1979) Patterns of inflammation in prostatic hyperplasia: A histologic and bacteriologic study. J Urol 121:755
Krastanova LJ, Addonizio JC (1981) Carcinosarcoma of the prostate. Urology 18:85
Kurth KH, Altwein JE, Skoluda D, Hohenfellner R (1977) Follow-up of irradiated prostatic carcinoma by aspiration biopsy. J Urol 117:615
Lederer B, Mikuz G, Gütter W, Zur Nedden G (1972) Zytophotometrische Untersuchungen von Tumoren des Übergangsepithels der Harnblase. Vergleich zytophotometrischer Untersuchungsergebnisse mit dem histologischen Grading. Beitr Path 147:379
Leistenschneider W (1981) Prostatakarzinom: Zytologie. Verhdlg Dtsch Ges Urol. 32:17
Leistenschneider W (1981) Zytologie und Zellkern-DNS-Analyse durch Zytophotometrie beim behandelten Prostatakarzinom und ihre Bedeutung für die Beurteilung von Therapieeffekt und Prognose. Habilitationsschrift, Berlin, 1981
Leistenschneider W (1982) Nucleic acid determination as therapy monitoring of prostate cancer. In: Jacobi GH, Hohenfellner R Prostate Cancer. Williams and Wilkins, Baltimore London
Leistenschneider W (1982) Zytodiagnostik. In: Hohenfellner R, Zingg EJ: Urologie in Klinik und Praxis, Bd 1. Thieme, Stuttgart New York
Leistenschneider W, Nagel R (1978a) Komplikationen bei transrektaler Stanz- und Feinnadelbiopsie. Therapiewoche 28:1963
Leistenschneider W, Nagel R (1978b) Prostatakarzinom: Wert der zytologischen Verlaufskontrollen unter verschiedenen Therapiebedingungen. Verhdlg Dtsch Ges Urol 29:335
Leistenschneider W, Nagel R (1978c) Zytologische Diagnose und Klassifizierung entzündlicher Prostataerkrankungen. Akt Urol 9:185
Leistenschneider W, Nagel R (1979a) Bestimmung des DNS-Gehaltes des behandelten Prostatakarzinoms mit Scanning-Zytophotometrie. Verhdlg Dtsch Ges Urol 30:399
Leistenschneider W, Nagel R (1979b) The cytologic differentiation of prostatitis. Path Res Pract 165:429
Leistenschneider W, Nagel R (1979c) Zellkern-DNS-Analyse am unbehandelten und behandelten Prostatakarzinom mit Scanning-Einzelzellzytophotometrie. Akt Urol 10:353
Leistenschneider W, Nagel R (1980a) Untersuchungen zum Zellkern-DNS-Gehalt von Prostataadenomen und -karzinomen. Verhdlg Dtsch Ges Urol 31:364
Leistenschneider W, Nagel R (1980b) Cytological and DNA-cytophotometric monitoring of the effect of therapy in conservatively treated prostatic carcinomas. Scand J Urol Nephrol 55:197 (Suppl)
Leistenschneider W, Nagel R (1980c) Estracyt therapy of advanced prostatic cancer with special

reference to control of therapy with cytology and DNA-cytophotometry. Europ Urol 6:111

Leistenschneider W, Nagel R (1980d) Zytologisches Regressionsgrading und seine prognostische Bedeutung beim konservativ behandelten Prostatakarzinom. Akt Urol 11:263

Leistenschneider W, Nagel R (1981) Zytologische Klassifizierung der Prostatitis. 2. Internationale Arbeitstagung „Chronische Prostatitis", Bad Nauheim, 27.–28.11.1981

Leistenschneider W, Nagel R (1983a) Einzelzellzytophotometrische Zellkern-DNS. Analysen beim behandelten, entdifferenzierten Prostatakarzinom und ihre klinische Bedeutung. Urologe A 22:157

Leistenschneider W, Nagel R (1983b) Estracyt-Therapie beim fortgeschrittenen Prostatakarzinom: Ergebnisse einer klinisch und zytologisch kontrollierten, prospektiven Studie. Akt Urol 14:127

Leuchtenberger C, Leuchtenberger R, Davis AM (1954) A microspectrophotometric study of the desoxyribose nucleic acid (DNA) content in cells of normal and malignant human tissues. Amer J Path 30:65

Levi PE, Cooper EH, Anderson CK, Path MC, Williams RE (1969) Analysis of DNA content, nuclear size and cell proliferation of transitional cell carcinoma in man. Cancer 23:1074

Maksimović P, Lübke W, Nagel R (1971) Erfahrungen mit der perinealen und transrektalen Biopsie der Prostata. Akt Urol 2:9

Meares EM, Stamey TA (1972) The diagnosis and management of bacterial prostatitis. Brit J Urol 44:175

Melchior J, Valk WL, Foret JD, Mebust WK (1974) The prostate in leukemia: Evaluation and review of literature. J Urol 111:647

Melicow MM, Pelton Th, Fish GW (1943) Sarcoma of the prostate gland: Review of the literature; table of classification; report of four cases. J Urol 49:675

Melograna F, Oertel YC, Kwart AM (1982) Prospective controlled assessment of fine-needle prostatic aspiration. Urology 19:47

Miller A, Seljelid R (1971) Cellular atypia in the prostate. Scand J Urol Nephrol 5:17

Mostofi FK (1975) Grading of prostatic carcinoma. Cancer Chemother Rep 59:111

Mostofi FK, Price EB (1973) Tumors of the male genital system. Atlas of tumor pathology. Sec Ser Fasc 8, AFIP, Washington DC

Mostofi FK, Sobin LH (1977) Histological typing of testis-tumours. International histological classification of tumours, No. 16, WHO 1977, Geneva

Müller HA, Altenähr E, Böcking A, Dhom G, Faul P, Göttinger H, Helpap B, Hohbach Ch, Kastendieck H, Leistenschneider W (1980) Über Klassifikation und Grading des Prostatakarzinoms. (Bericht einer Pathologisch-Urologischen Arbeitsgruppe „Prostatacarcinom", Würzburg, Berlin, Freiburg, Homburg, Memmingen, München, Bonn, Hamburg, Homburg, Berlin). Verhdlg Dtsch Ges Path 64:609

Müller HA, Wünsch PH (1981) Features of prostatic sarcomas in combined aspiration and punch biopsies. Acta Cytol 25:482

Mulholland SW (1931) A study of prostatic secretion and its relation to malignancy. Proc Staff Meeting Mayo Clin 6:169

Nagel R, Leistenschneider W (1982) Unspezifische Entzündungen der Harnröhre und männlichen Adnexe. In: Hohenfellner R, Zingg EJ: Urologie in Klinik und Praxis, Bd 1. Thieme, Stuttgart New York

Narayama AS, Loening S, Weimar GW, Culp DA (1978) Sarcoma of the bladder and prostate. J Urol 119:72

Nielsen ML, Asnaes S, Hattel T (1973) Inflammatory changes in the non-infected prostate gland. A clinical, morphological and histological investigation. J Urol 110:423

O'Shaugnessy EJ, Parrino PS, White JD (1956) Chronic prostatitis – fact or fiction? JAMA 160:540

Papanicolaou GN (1954) Atlas of exfoliative cytology. Harvard Univ Press Cambridge, Mass

Peck S (1960) Needle biopsy of prostate. J Urol 83:176

Puigvert A, Jiménez F, Manavella JN (1975) Implantation perinéale néoplasique après la biopsie de la prostate. Urol Int 30:305

Rost A, Riedel B, Pust R (1975) Die zytologische Kontrolle der Effizienz der antiandrogenen Therapie des Prostatakarzinoms. Urol Int 30:245

Rothkopf M (1966) Unsere Erfahrungen mit der transrektalen Prostatastanzbiopsie. Zschr Urol 18:273

Ruch F (1970) Principles and some applications of cytofluorometry. In: Introduction to quantitative cytochemistry, Vol 2. Academic Press, New York

Ruch F, Leemann U (1973) Cytofluorometry. In: Neuhoff V Micromethods in molecular biology. Springer, Berlin Heidelberg New York

Ruch F (1979) Zytofluorometrie. In: Witte S, Ruch F Moderne Untersuchungsmethoden in der Zytologie, 2. Auflage. Gerhard Witzstrock, Baden-Baden Köln New York

Sandritter W (1952) Über den Nucleinsäuregehalt in verschiedenen Tumoren. Frankf Z Path 63: 423

Schuppler J (1971) Malignant neurolemmoma of prostate gland. J Urol 106:903

Seppelt U, Sprenger E (1981) Zellkern-DNS-Analysen durch Einzelzell-Fluoreszenz-Zytophotometrie an Prostatakarzinomen vor und während der Therapie. Verhdlg Dtsch Ges Urol 32:68

Sewell RA, Braren V, Wilson SK, Rhamy RK (1975) Extended biopsy follow up after full course radiation for resectable prostatic carcinoma. J Urol 113:371

Sika JV, Lindquist HD (1963) Relationship of needle biopsy diagnosis of prostate to clinical signs of prostatic cancer: An evaluation of 300 cases. J Urol 89:737

Smith BH, Dehner LP (1972) Sarcoma of the prostate gland. Amer J Clin Path 58:43

Soost HJ (1978) Lehrbuch der klinischen Zytodiagnostik. Thieme, Stuttgart

Spieler P, Gloor F, Egle N, Bandhauer K (1976) Cytological findings in transrectal aspiration biopsy on hormone- and radio-treated carcinoma of the prostate. Virchows Arch A (Pathol Anat) 372:149

Sprenger E (1979) Anwendungsmöglichkeiten der Durchflußphotometrie in der zytologischen Diagnostik. In: Witte S, Ruch F Moderne Untersuchungsmethoden in der Zytologie, 2. Auflage. Gerhard Witzstrock, Baden-Baden Köln New York

Sprenger E, Volk L, Michaelis WE (1974) Die Aussagekraft der Zellkern-DNS-Bestimmung bei der Diagnostik des Prostatakarzinoms. Beitr Path 153:370

Sprenger E, Michaelis WE, Vogt-Schaden M, Otto C (1976) The significance of DNA flow through fluorescence cytophotometry for the diagnosis of prostate carcinoma. Beitr Path 159:292

Staehler W, Ziegler H, Völter D (1975) Zytodiagnostik der Prostata. Grundriß und Atlas. Schattauer, Stuttgart New York

Takahashi M (1981) Color atlas of cancer cytology. Thieme, Stuttgart New York

Tanke HJ, Ploem JS, Jonas U (1982) Kombinierte Durchflußzytophotometrie und Bildanalyse zur automatisierten Zytologie von Blasenepithel und Prostata. Akt Urol 13:109

Tannenbaum M (1975) Sarcomas of the prostate gland. Urology 5:810

Thackray AC, Crane WAJ (1976) Seminoma. In: Pugh RCB Pathology of the testis. Blackwell Scientific Publications, Oxford London Edinburgh Melbourne

Tümmers H, Weissbach L (1975) Komplikationen nach transrektaler Stanzbiopsie der Prostata. Vortrag auf der 16. Tagung d. Südwestdeutschen Ges f. Urologie, Reutlingen 1975

Voeth Ch, Droese M, Steuer G (1978) Erfahrungen mit dem zytologischen Grading beim Prostatakarzinom. Urologe A 17:367

von Sengbusch G, Hugemann B (1974) A fluorescence microscope attachment for flow-through cytofluorometry. Exptl Cell Res 86:53

Wajsman J, Mott LJ (1978) Pathology of neoplasms of the prostate gland. In: Skinner DG, de Kernion JB Genitourinary cancer. WB Saunders, Philadelphia, London, Toronto

Wendel RG, Evans AT (1967) Complications of punch biopsy of the prostate gland. J Urol 97:122

Wullstein HK, Müller HA (1973) Zur zytologischen Diagnose der granulomatösen Prostatitis im Prostata-Aspirat. Verhdlg Dtsch Ges Path 57:333

Zetterberg A, Esposti PL (1976) Cytophotometric DNA-analysis of aspirated cells from prostatic carcinoma. Acta Cytol 20:46

Zimmermann A, Schauer A, Truss F (1979) Automatisierte Zellkern-DNS-Bestimmung zur Diagnostik des Prostatakarzinoms. Akt Urol 10:347

Zincke H, Campbell JT, Utz DC, Farrow GM, Anderson MJ (1973) Confidence in the negative transrectal needle biopsy. Surg Gynecol Obstet 136:78

Subject Index

MIX
Papier aus verantwortungsvollen Quellen
Paper from responsible sources
FSC® C105338

If you have any concerns about our products,
you can contact us on
ProductSafety@springernature.com

In case Publisher is established outside the EU,
the EU authorized representative is:
Springer Nature Customer Service Center GmbH
Europaplatz 3, 69115 Heidelberg, Germany

Printed by Libri Plureos GmbH
in Hamburg, Germany